The Texas Tortoise

ANIMAL NATURAL HISTORY SERIES
Victor H. Hutchison, *General Editor*

Also by Francis L. Rose
(and Russell W. Strandtmann) *Wildflowers of the Llano Estacado*
   (Lubbock, Texas, 1986)

Also by Frank W. Judd
(co-editor with John W. Tunnell, Jr.) *The Laguna Madre of Texas
   and Tamaulipas* (College Station, Texas, 2001)

Library of Congress Cataloging-in-Publication Data
Rose, Francis L.
 The Texas tortoise : a natural history / Francis L. Rose
and Frank W. Judd.
    pages cm
 Includes bibliographical references and index.
 ISBN 978-0-8061-4451-1 (hardcover : alk. paper)
 1. Texas tortoise.   I. Judd, Frank W.   II. Title.
 QL666.C584R675 2014
 597.92'46—dc23
                                2013044689

The Texas Tortoise: A Natural History is Volume 13 in the Animal
Natural History Series.

The paper in this book meets the guidelines for permanence and
durability of the Committee on Production Guidelines for Book
Longevity of the Council on Library Resources, Inc. ∞

1 2 3 4 5 6 7 8 9 10

Contents

Illustrations

Color Plates

Figures

Maps

Tables

Preface

Tottering, two-legged, stilt-gaited beasts proportions all awry . . . each
sense dulled by mental acuity. Reason in place of a good nose. Logic
instead of a tail. Faith instead of certain knowledge of instinct. . . .
Superstition instead of a shell.

Reverend Gilbert White, Humans, 1789

We receive a series of general questions when someone discovers that we study
Texas tortoises. The most frequent question is, "How old do they get?" What
the person means to ask is, "How old is the oldest one that you know about?"
The next most frequent question is, "How does a turtle differ from a tortoise?"
On the surface, these questions are simple and direct, but truthful answers will
provide more information than the questioner may have had in mind.

We began our studies in the late 1960s. Both of us had yearlong stays at the
Florida Museum of Natural History at the University of Florida in Gainesville.
Walter Auffenberg (fig. P.1) was director of the Natural Sciences Department
at that time, and William Weaver was a doctoral student. Together they pio-
neered population biology studies on the Texas tortoise. Ironically, it was Don-
ald Tinkle, another herpetologist, who pointed out to Walter Auffenberg that
lomas along the Texas coast had substantial populations of large tortoises. We
say ironic because at that time Don Tinkle was with Texas Tech University,
where one of us (FLR) began a teaching and research career in 1966, and the
other (FWJ) obtained a doctoral degree in 1973.

Of the five species of North American land tortoises, Texas tortoises have
been the most neglected by biologists and wildlife personnel charged with their
welfare. Although their geographic range has shrunk considerably because of
land alteration (orchards, truck farming, ranching), there are large blocks of
ranchland where tortoises are somewhat protected. Laws and state regulations

Figure P.1. Walter Auffenberg (1928–2004): Paleontologist, herpetologist, educator, artist, doyen, and the person who initiated fundamental work on tortoises, fossil and living. Courtesy of Kurt Auffenberg.

enacted to protect tortoises are only as effective as the level at which they are enforced. No gold stars will be given to those saddled with that challenging responsibility in Texas. In addition, the status of the tortoise in Mexico is virtually unknown, and there are no indications as to population trends in its vast range in that country. More roads and vehicles enter the tortoise's realm each year and they take a toll, and not just along highways. Many tortoises are killed inadvertently every year on ranch roads because of their propensity to use roadways for foraging, and often, while they hide in the grass, they lose the battle with vehicles.

We know that some captive tortoises attain ages in excess of 90 years, and they are no larger than some we have observed in the wild. Mr. A. H. George found a male tortoise known as G4 in 1927. The tortoise was given to Mr. Carlton McQueen in 1977. In November 1996 its carapace was 210 mm (8.3 in) long, and today (September 2010) its carapace is 218 mm (8.6 in) long, near maximum size for the species. Undoubtedly, females have shorter life spans, for their fate is to suffer from the negative effects of egg production, which stunts their growth and thins their bones to such a degree that they are more easily preyed upon.

Standard (common) names have utility, but problems arise when there are local identical or similar names for different organisms. Carolus Linnaeus solved this problem when he proposed the system of binomial nomenclature in the 1730s, wherein each organism type was assigned a genus and species name. So, the names cooter, slider, terrapin, turtle, and tortoise conjure up images of some form of chelonian, that is, a turtle. Turtles can be defined as various aquatic or terrestrial reptiles assigned to the order Testudines (or Chelonia) and having horny, toothless jaws, and a bony or leathery shell into which the head, limbs, and tail can be withdrawn. Tortoises are defined as any of various terrestrial turtles, especially those in the family Testudinidae, which are characterized by thick, club-like (elephantine) hind feet, generally with four unwebbed toes and no more than two phalanges (toe bones) in each digit; bony scales (osteoderms) on the forelimbs; and often a high, rounded carapace. So, tortoises are turtles belonging to the family Testudinidae, and they are generally associated with hot, dry environments, but several species inhabit humid forest swamps. They range in size from the Australian speckled padlopers (*Homopus signatus*) at 96 mm (3.7 in) and 140 grams (0.3 lb) to Galapagos tortoises (*Geochelone elephantopus*) at 1340 mm (53 in) and 263,000 grams (579 lb).

We undertook writing this book to elucidate and consolidate what is known about the Texas tortoise. It is probably worth noting that just about all we know about the ecology and life history of this tortoise is based on studies conducted in thorn-shrub communities of Cameron County on lomas east of Brownsville (Auffenberg and Weaver 1969), studies at the Laguna Atascosa National Wildlife Refuge (Bury and Smith 1986), our studies on the Yturria and Reed Ranches in Cameron County, and on several studies conducted on the Chaparral Management Area in Dimmit and La Salle Counties (Hellgren et al. 2000; Kazmaier et al. 2001a, 2001c). Little has been reported about the biology of this tortoise in Mexico generally or in oak-savanna communities in both Mexico and Texas where it is known to occur, albeit at lower densities. When we started our studies, Global Positioning Systems were not available, nor were the computers and data-crunching programs that are so prevalent today. During this time, karyotype and monoclonal antibody comparisons, starch and polyacrylamide gel electrophoresis, immunological techniques, and DNA annealing came and went as prevailing methodologies for elucidating organismal relationships, as did various other uses of DNA. Many researchers developed stellar careers based on one or more of these techniques, but today, vast volumes of their work have passed into the woodpile of irrelevance. More importantly, during this time the study of biology passed from the knowledge stage to the information stage, for it is information that is spewed forth from computer programs, not knowledge. Biologists rarely ruminate on data today but instead leave that task to sophisticated computer programs. T. S. Eliot asked in 1934, "Where is the wisdom we have lost in knowledge? And where is the knowledge we have lost in information?" We rest heavily today in the belly of the information beast, with data outlets so numerous that syntheses are becoming less probable even as global changes threaten the existence of many species: tortoises of the world will rue this state of affairs. More importantly, from where will there emerge a new generation of natural history biologists and taxonomists who are not only willing to suffer through the rigors of data gathering but who can articulate the intrinsic value of organisms in a public forum, and who carry the intellectual weight to do so?

In particular, we wish to convey how this wonderful creature, the Texas tortoise, functions in and interacts with its environment. Along the way we discuss its relationships to fossil and extant tortoises (chapter 1), its environment (chapter 2), and those things that it eats and those things that eat it (chapter 3). We may have delved a bit too much into morphology (chapter 4), but what is a tortoise if not an interesting morphological type? How tortoises grow and reproduce is covered in chapter 5. We know very little about sensory modalities (chapter 6) but quite a bit about temperature and water relationships (chapter 7), behavior (chapter 8), and population ecology (chapter 9). We are familiar with many facets of tortoise survival (chapter 10) and have dark and foreboding concerns about its future protection (chapter 11). Also in chapter 11, we provide suggestions for tortoise care and maintenance, and this burdens us with a dilemma, for on one hand we wish to admonish those who pick up tortoises along roads to leave them alone, but we also feel compelled to offer tips on how to care for them in captivity. We speculate (chapter 12) about the significance of the Texas tortoise as a keystone species and about how its general ecological fate is intertwined with other species associated with South Texas and northeastern Mexico thorn-shrub communities. We hope that this book will be helpful to many who share an interest in these fascinating creatures, and that it will shed light on possible new research paths. When dinosaurs were begging their new constructs not to fly away, tortoises, along with crocodilians, remained loyal to their kin and survived many millions of years. Now, some find themselves, like hormone-driven lemmings, on an accelerated trajectory nearing the dark edge of their demise: extinction may be inevitable, but we are not required to like it.

There are some redundancies in the book, and these are by design. Books such as this about a specific organism are not novels that follow circuitous paths to an inevitable conclusion. They are used to inform on specific subjects, and therefore, for completeness we have chosen to include redundancies in appropriate sections.

Acknowledgments

We have benefited from and enjoyed the help and encouragement of many people associated with our work. Walter Auffenberg provided the original impetus that led us down a long and circuitous path. For general and logistical support we thank the biological sciences or biology departments at Texas Tech University, Texas State University, and the University of Texas–Pan American. Bruce Bury, through the U.S. Fish and Wildlife Service, provided funding for the project, and special thanks are due to the 77 volunteers sponsored by EARTHWATCH (Center for Field Research), who enabled us to monitor egg production. The International Center for Arid and Semiarid Land Studies and the World Wildlife Fund provided support funds, as did Larry McKinney through the Texas Parks and Wildlife Department. We owe special thanks to Frank D. Yturria, the Frank D. Yturria Trust, and Wallace Reed for allowing us access to their land. Linda Irwin, Jimmy Kenney, Gary Matney, Lew Milner, and Leland Parks braved unsavory elements to physically develop the Yturria or Reed Ranch grids and helped with monitoring at various times. Donna Precure-Rose, Jane Judd, Amy Spangler, Dede Armentrout, Art Laffer, Irmaruth Mahlstedt Rose, and John and Terris Rose helped with surveys and provided help in several other ways that enhanced the work. Kenny King coordinated and handled a suite of volunteers associated with EARTHWATCH. Veterinarians, especially Elliott Jacobson and Jeff Jorgenson, and M. L. and E. M. Feldman, Angela Jordan, and Sally and Dan Nowlin provided advice and help with tortoises in need. Don Anders and John Senter produced several photographic images and Tom Clendennon helped with copying. Roxana Tuff and Michael Nickell provided artwork. We benefited greatly from the help and counsel of Michael Forstner, Chris Nice, and Jim and Eva Ott, and especially Thomas Simpson. For help in obtaining and using images or literature we thank Roy Averill-Murray, Jeff Bundy, Michael Dloogatch, Dan Guthrie, Brent Palmer, Richard Seigel, David Synatzske, and Daniel Walker. Robert Lonard has been most helpful for many years regarding information on and identification of plants. We benefited greatly from the accumulated Texas tortoise knowledge generously provided by Carlton McQueen. Annie Simpson was extremely helpful in developing image labeling. James Dixon and Richard Franz provided helpful guidance in manuscript review. Jeremy Wicker provided clarifications on several issues whenever he was asked. We thank the staff of the Oklahoma University Press, notably Jay Dew, Emily Jerman, and Victor Hutchison, and we thank our copyeditor, Laurel Anderton; all made significant contributions while shepherding the manuscript to final form.

We thank the American Museum of Natural History for permission to use the image of Louis Agassiz, the Society for the Study of Amphibians and Reptiles for permission to use the two images of the plastral hinge of Texas tortoises, the Texas Parks and Wildlife Department (Chaparral Wildlife Management Area) for the image of fire damage in Texas tortoise habitat, Robert and Linda Mitchell for use of the image of the yellow form of the Texas tortoise, and the Texas State Historical Association for permission to use the map of Berlandier's travels and quote from *Journey to Mexico during the Years 1826 to 1834*. We thank Viking Penguin, a division of Penguin Group (USA), Inc., for copyright permission to use the quote from *The Log from the Sea of Cortez* by John Steinbeck, copyright 1941, 1951 by John Steinbeck and Edward F. Ricketts, Jr., with copyright renewal in 1969 by John Steinbeck and Edward F. Ricketts, Jr. The excerpt from *American Nomads* (2003), by Richard Grant, is used with permission of Grove/Atlantic, Inc. Cornell University Press granted permission to use the quoted material by Dr. Archie Carr found in *Handbook of Turtles: The Turtles of the United States, Canada, and Baja California* (1952). The Florida Museum of Natural History gave permission to use the image copied from the *Bulletin of the Florida Museum of Natural History, Biological Sciences Series*. Kurt Auffenberg provided the image of his father.

Finally, we acknowledge the help and support of our wives, Donna Precure-Rose and Jane Judd. We would be remiss in failing to point out the many years they have supported our efforts, enduring hardships associated with data collecting, and indulging our research activities, triumphs, and failures. Donna's time and effort in caring for injured tortoises and nursing hatchlings is a debt that cannot be paid. Her kindness to those creatures needing help cannot be measured, but is acknowledged. Jane has been involved many times for so many years, giving of her time to accommodate not only our efforts, but also the students and volunteers who have worked on projects.

The Texas Tortoise

Introduction

Indeed, in America generally, the traveller who would behold the finest
landscapes, must seek them not by the railroad, nor by the steamboat, nor
by the stage-coach, nor in his private carriage, nor yet even on horseback—
but on foot. He must walk, he must leap ravines, he must risk his neck
among precipices, or he must leave unseen the truest, the richest, and
most unspeakable glories of the land.
                    —Edgar Allan Poe, "Morning on the Wissahiccon"

Upon its discovery, explorers poured into the New World for various reasons,
many to claim its riches for self and church. A handful of intrepid souls braved
harsh environmental conditions and hostile inhabitants—their human coun-
terparts being lumped in the latter category—and provided an early plethora
of specimens craved by the frenzied taxonomists of Europe and America.
Jean Louis Berlandier was such a New World traveler. He not only garnered
specimens—including the first specimens of what is now known as the Texas
tortoise—but he made voluminous notes of his travels. He was the first naturalist to cover much of the southern portion of what was to become the state of
Texas.

Berlandier was born in about 1805 to parents of modest means in France,
where he educated himself in Latin and Greek and worked in a pharmaceutical
house, probably as an apprentice. Botany seemed to be his forte and he pursued
his interests at the University of Geneva. His exceptional talent, like that of
a young Charles Darwin, was noticed and facilitated by several mentors. Berlandier was sent to Mexico by Professor Augustin Pyramus de Candolle of the
Geneva Academy in Switzerland to collect natural history specimens and make
notes on the country. As a botanist, de Candolle had an imposing reputation
and he arranged for Berlandier to be attached to the Mexican Boundary Com-

mission as botanist and zoologist. Berlandier was thus required to collect a wide variety of specimens, but as a consequence of his training and his sponsor's interest, plant specimens clearly dominated.

In December 1826, Berlandier arrived in Veracruz, Mexico. He joined the Mexican Boundary Commission and made his first excursion into what would become Texas, entering at Laredo in February 1828. He made plant collections in the San Antonio area and east to what is now Anderson in Grimes County, Texas, where he endured the debilitating effects of malaria. He returned to Matamoros but soon went back to San Antonio, where he participated in an expedition to silver mines along the San Saba River. Later in the year he visited Goliad as part of a military expedition to quell an uprising against the presidio commander. He returned to Matamoros when the Boundary Commission was inactivated in November 1829 (map 1). Note that this segment of his life included only 21 months.

On his return to Matamoros he married a local woman, became a physician, and was well liked and respected for his caring manner. He made several other excursions, including one to Goliad in 1834, but became entangled in the Mexican War, where he served as an interpreter and oversaw hospitals in Matamoros. He drowned in 1851 attempting to cross the San Fernando River, at a point 137 km (85 mi) south-southwest of Matamoros.

Unfortunately, de Candolle, though once Berlandier's strongest advocate, ultimately became his greatest detractor (Geiser 1948). This winnowing of respect stemmed from what de Candolle interpreted as a lack of professional courtesy from Berlandier and his disappointment in the quality and quantity of specimens Berlandier sent him. Asa Gray, who lived from 1810 to 1888 and was the father of plant geography, also had low esteem for Berlandier (Dupree 1968) and considered him a "dishonest knave," but it was Gray who received plant collections from Berlandier and who placed them in the mainstream of botanical inquiry.

Louis Agassiz (fig. I.1) was a towering scientific presence in the nineteenth century and described what was to become the Texas tortoise as *Xerobates berlandieri*. In that description he commented: "Collected by the late Mr. Berlandier, a zealous French naturalist, to whom we are indebted for much of what we know of the natural history of northern Mexico." This praise of Berlandier was as much a taunt at Gray as anything else because both men had agreed to disagree; they were contemporaries at Harvard, where Gray fostered what is now the Gray Herbarium and Agassiz established the Museum of Comparative Zoology. Gray saw the value of Darwin's views regarding species and their origins; Agassiz was a staunch creationist and in his earlier years recognized no genetic continuity between individuals of a species, although he seems to have modified this belief toward the end of his career. Personal opinions, biases, and religious beliefs do not necessarily make bad taxonomy, but they do lead to bad science. In spite of his personal beliefs, Agassiz was a staunch supporter of sci-

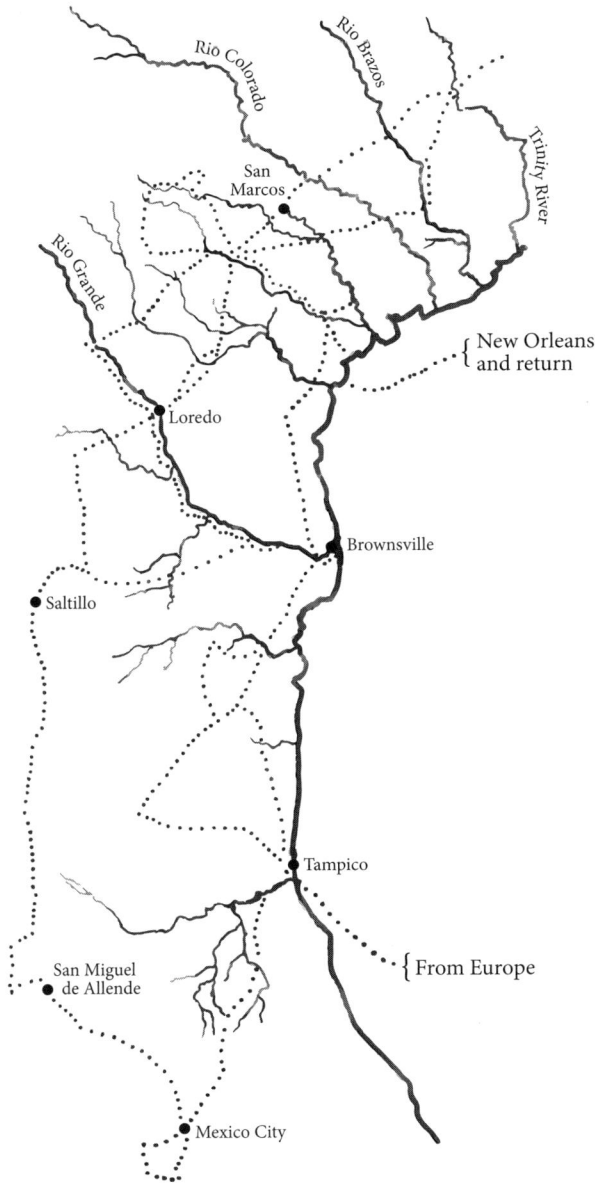

Map 1. Jean Louis Berlandier's extensive travels in Mexico and Texas. The purpose of the map is to show the degree of potential hardships he encountered during his travels, while others, who were so critical of his efforts, remained cloistered in ivory towers. Condensed from two maps in Ohlendorf et al. (1980), vol. 1. Original cartography by David M. Ridner. Used with permission of the Texas State Historical Association.

Figure I.1. Louis Agassiz, who described the Texas tortoise, lecturing at Harvard University. Image no. 28001, used with permission of the American Museum of Natural History Library.

ence, and he and Gray are eminent figures in the historical scientific landscape. It is perhaps paradoxical that both Gray and Agassiz started out as physicians, and Berlandier went out as one.

While Berlandier was braving the privations and hardships of an untamed land, both de Candolle and Gray remained securely nested in their hallowed halls, describing and assigning names to organisms. One can only wonder, with their encyclopedic minds, what the pace of discovery might have been had they personally taken up the challenges of exploration and collection. Perhaps Gray and de Candolle might have more intimately appreciated the joys of outdoor living at its best, beset with droughts, floods, scorching heat, cold, rain, mosquito-borne diseases, venomous snakes, chiggers, fleas, ticks, scorpions, giant centipedes, cactus spines, vermin, boils, hostile natives, and primitive transportation with limited storage and shipping venues.

Steinbeck (1995) referred to biologists cloistered in deep recesses of moldy buildings as "Dry Balls." But Berlandier was no Dry Ball, and his designation as "a scapegoat in the history of botanical exploration in the Southwest" is unwarranted and indefensible, as pointed out by Geiser (1948). There has been some disagreement about the term "Dry Balls," but apparently it refers to individuals who are sterile because of excessive onanism.

Louis Agassiz, in his description of *Xerobates berlandieri* (Agassiz 1857), relied on specimens sent to him by U.S. National Museum staff. These specimens were acquired in 1853 from Jean Louis Berlandier's widow in Matamoros, Mexico, by then-Lieutenant Darius N. Couch, U.S. Army, and brought to the U.S. National Museum.

Berlandier's original manuscript describing his travels and collecting for de Candolle and the Mexican Boundary Commission was among the manuscripts obtained from his widow by Lt. Couch. It was written in French on about 1,500 pages and titled "Voyage au mexique par Louis Berlandier pendant les années 1826 á 1834"; it is now housed in the Library of Congress. The manuscript was translated into English by Ohlendorf et al. (1980) and was published in two volumes by the Texas State Historical Association in cooperation with the Center for Studies in Texas History, University of Texas at Austin.

Examination of the two volumes reveals that Berlandier makes only two references to tortoises in his description of his travels with the Boundary Commission. Between Santa Teresa and San Fernando, Tamaulipas, Berlandier described finding two forms of terrestrial turtles. Translators of the manuscript apparently accepted that Berlandier referred to two valid species. Because Berlandier stated that two forms were common on both banks of the Rio Bravo (also known as the Rio Grande), they pointed out, in a footnote, that only two species of terrestrial turtles occurred near the Rio Grande, *Gopherus berlandieri* and the ornate box turtle, *Terrapene ornata*. We are convinced that Berlandier was not describing *Terrapene ornata*. Rather, he was probably describing two ontogenetic stages or two geographic variants of the Texas tortoise. His descriptions are short, and it is necessary to read them to understand the reasoning behind our conclusion. Consequently, we include the entire description: "Two species of terrestrial turtles were found in these low-lying regions and they are common on both banks of the Rio Bravo. One of them (*Testudo tuberculata* B. mss.) is remarkable for two rounded tubercles under the neck. It often serves as food for the military of the presidios when they travel in the wilderness. The other species of these turtles, which I have designated as *Testudo bicolor*, is very small and, like the preceding, has at the anterior part of the plastron two teeth, or prolongations. These are so long in the tuberculose turtle that they serve to support its neck and head."

Clearly, Berlandier knew the difference between *Gopherus berlandieri* and *Terrapene ornata*, for he remarked that he found a terrestrial turtle "of the kind whose plastron is divided transversally by a hinge into two parts: one part could be called pectoral and the other abdominal." And, he stated in his descriptions that both kinds had prolongations on the anterior part of the plastron. These projections are lacking in box turtles. The form he called *Testudo tuberculata* was likely based on a mature male Texas tortoise with well-developed subdentary glands (Smith and Brown 1946; Rose et al. 1969; Rose 1970). The small form that Berlandier called *Testudo bicolor* was likely a young Texas tortoise (plate 1). Agassiz (1857) wrote that the median and costal scales of young tortoises have a yellow center.

Auffenberg and Weaver (1969) pointed out that hatchlings and young have distinctly bicolored shells, with the centers of the vertebral and costal scutes of the carapace being a creamy white and the surrounding area being dark brown or black. The centers of the scutes may also be yellow instead of creamy white. As they age, tortoises transition from black and yellow to brown and yellow to a uniform brown or "horn" color, and Auffenberg and Weaver (1969) noted that all old tortoises are a light brown or horn color, which we also found to be true. It is clear that Berlandier was describing a Texas tortoise because he mentioned gular projections but did not mention a hinged plastron. A second interpretation is that Berlandier described the larger, more horn-colored adults occurring along his coastal route and the smaller, more brightly colored individuals along his more inland routes. The only other reference Berlandier made to tortoises was to state that they were often used as food by the military when traveling from one presidio to another.

Tortoises are among our oldest extant lineages and have enjoyed a splendid and diverse history. They inhabit all kinds of habitats, from wet Amazonian forests to the Mojave Desert, one of the hottest and driest environments on earth. That they did so well for so long is perplexing at times, as they are slow, generally require extended time to reach sexual maturity, do not actively defend themselves against predators, and are readily eaten by all manner of creatures including human beings. The plight of these wonderful creatures is saddening and sickening, for the present world is not their world, and their numbers have dwindled such that most species are in need of protection from exploitation and habitat destruction. Nowhere has their history been so eloquently and succinctly captured as by Archie Carr in the *Handbook of Turtles* (1952):

> The Cenozoic came, and with it progressive drought, and the turtles
> joined the great hegira of swamp and forest animals to steppe and prai-
> rie, and watched again as the mammals rose to heights of evolutionary

frenzy reminiscent of the dinosaurs in their day, and swept across the grasslands in an endless cavalcade of restless, warm-blooded types. Turtles went with them, as tortoises now, with high shells and columnar, elephantine feet, but always making as few compromises as possible with the new environment, for by now their architecture and their philosophy had been proved by the eons; and there is no wonder that they just kept on watching as Eohippus begat Man o' War and a mob of irresponsible and shifty-eyed little shrews swarmed down out of the trees to chip at stones, and fidget around fires, and build atom bombs.

One must wonder, as the curtain falls, who will sing for these magnificent creatures now that the giants upon whose shoulders we so often stood have left the stage.

1

General Description and Phylogenetic Relationships

It proceeds more by accident and confusion. It turns back on itself and takes triumphant detours up blind alleys. What appears, with hindsight, to be a straightforward process of cause and effect is always fraught with exceptions, paradoxes and strange offshoots. The present is gone before we can make sense of it, and the future is made up of distorted echoes from the past.

—Richard Grant, *American Nomads*

Before we can discuss the Texas tortoise, we have to define what exactly a tortoise is. Early on in the discovery revolution, from the mid-1750s to the mid-1850s, there was a rush to apply a name to one or more specimens of every animal and plant to indicate species status; however, at that time there was not agreement as to what a species actually was (today we have more than 20 definitions of the word). Carolus Linnaeus, a Swedish biologist, brought focus to the plethora of new descriptions (and here we want to point out that the word "description" was ambiguous at best) by proposing a system—binominal nomenclature—wherein each species described was given a genus name and a species designation, the couplet forming the species epithet. Although others, including Aristotle and Theophrastus, attempted to shoehorn organisms into slots, their systems lacked credence and staying power. (Aristotle's broad categories included "has blood" versus "has no blood.") Even today we struggle with the concepts of genus and species. There are "lumpers," who tend to combine loosely defined species together, decreasing diversity; and "splitters," who recognize species status based on what to many might be insignificant characters of morphology, behavior, physiology, or genetics. Ernst Mayr saw an opportunity to solidify the concept of a species and developed the biological species concept, which became the following generally accepted working definition: "Species are groups of inter-breeding natural populations that are reproductively isolated from other such groups." In shorter parlance, members of a species are reproductively compatible, but Mayr (1970) was quick to point out that species are also ecological units. Of course, it is easy to recognize that speciation is a process that requires some time if you are an animal, and the primary model (allopatric) of speciation demands geographic isolation. It is understandable, then, that some forms of animals that appear phenotypically, or physically, different from each other can freely interbreed when they are together because the ability to interbreed is a shared primitive character. The description of a new species does not require that the author(s) demonstrate reproductive incompatibility with recognized closely related species; demonstrating phenotypic differences implies it. As such, most species are described as morphological species with the inherent hope that differing morphologies signify genetic divergence, and the taxonomists are usually correct because they must jump through a series of defining hoops to make their case. As new techniques, and the math to handle these new and large data sets, came on board, taxonomic arrangements were confirmed or changed. But now we have come full circle in herpetology, and standard names of organisms are more stable than scientific ones. DNA analysis will unravel some relationships and cloud others, and we will do well to remember that such analysis is no less butt-numbing phenotypic drudgery than that of cloistered "Dry Balls" working immersed in cancer-inducing formalin and waving a pair of calipers. Today, one machine generates the data and another gives researchers options for what the results may mean and selects the most probable correct association. Some researchers have never seen a whole member of the group whose taxonomy they presume to elucidate, nor are they attuned to its natural history. The extant (living) North American tortoises, of which only five species are currently recognized, did not escape taxonomic conflict.

But before we turn our attention to tortoises, perhaps we will visit the terms *taxonomy* and *systematics* and why it is important that the specific epithet used as the name for an organism accurately reflects that organism's place in the taxonomic order. As the cell is the basic structure of protoplasmic organization, the species is the basic level of organization of groups of organisms. Without a species designation, there would be nothing to grasp, nothing to meaningfully order, nothing to conceptualize regarding the vast number of organisms inhabiting the thin, life-sustaining band of space around this spectacular planet where water can be liquid. All categories above the species level are somewhat arbitrary, but the species is the rock on which we anchor our organization. Without the species concept and category, organisms would be ordered and shoehorned into categories much the same way that a hardware store organizes its items for convenience of purchase. As human beings, we demand order and organization, for that is how we store information and how we are sure we are discussing the same item, whether it be a make of vehicle or a cluster of stars in a distant galaxy. In a selfish sense, we need to know what species we are dealing with relative to what that species might offer us in terms of breakthrough pharmaceuticals and antibiotics. It is not that we want to name every creature on earth; it is that we must try. Completing this task is, of course, a worthy goal that will never be accomplished.

Taxonomy is the orderly classification of organisms according to their presumed natural relationships; systematics can be defined the same way. As new methods came on line to answer questions of organization and relationships, the role of naming organisms remained with the taxonomists, and systematists emerged to explain how organisms were related and what their historical association (phylogenetic restoration) might have been. DNA analyses are at the heart of systematics today and can be extremely useful. For example, they allow researchers to compare data sets with fossil evidence to check the age of species separations. However, while we strain to create a definitive technique that will allow positive interpretations of relatedness, DNA techniques just inflame the battle of which genes to analyze, which out-groups to use in comparisons, and which personal and computer-generated interpretations carry the most weight. Taxonomists and systematists are, after all, human beings (a view not held by all cell and molecular biologists), and personal choices have to be made regarding data interpretation. Some systematists too often appear to be more concerned about the best data fit than they are about what the sum of the data tells us about relatedness. Then again, the glow of discovery and clarification would dim if an invariant technique arose.

Five species are recognized today in the genus *Gopherus*: *G. agassizii* (Agassiz desert tortoise), *G. berlandieri* (Texas tortoise), *G. flavomarginatus* (Bolson tortoise), *G. polyphemus* (gopher tortoise), and *G. morafkai*. The geographic ranges (map 2) of the five species do not overlap (they are allopatric). This is important when considering species status; however, their distributions supposedly overlapped (or, were sympatric) during part of the Pleistocene epoch. The standard name "desert tortoise" is somewhat misleading because the animal's successful paradigm had been in place "perhaps 12 million years before the formation of major regional deserts in North America" (Morafka and Berry 2002).

Duges (1888) stated that the then three described species were actually one species (*G. flavomarginatus* was not described by Legler until 1959, *G. lepidocephalus* [an earlier name for *G.* agassizii] not until 1989, and *G. morafkai* not until 2011), but that view was not given much credence. Auffenberg (1976), after reviewing morphology and fossils, concluded that four species were best represented in the genus *Gopherus*, and that *G. polyphemus* and *G. flavomarginatus* were more closely related than either was to *G. agassizii* or *G. berlandieri*. Bramble (1982) presented credible evidence that two genera should be recognized, with *G. polyphemus* and *G. flavomarginatus* being maintained in *Gopherus*, and *Scaptochelys* to include *G. agassizii* and *G. berlandieri*. Bour and Dubois (1984) pointed out that *Scaptochelys* was an earlier name (synonym) for

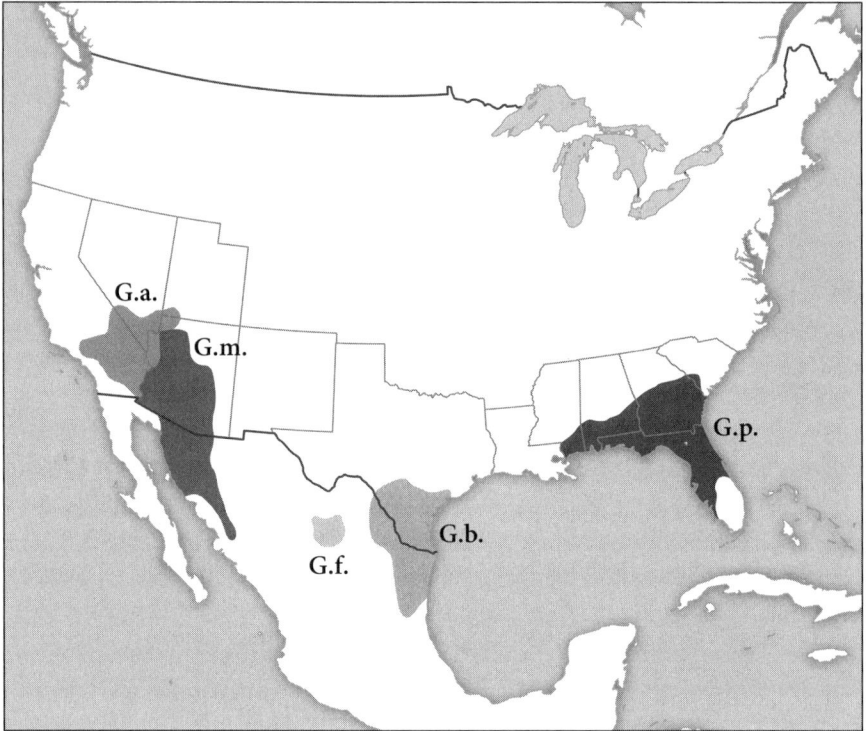

Map 2. General geographic ranges of the five living currently recognized species of the genus *Gopherus*. G. p. = *G. polyphemus*; G. b. = *G. berlandieri*; G. f. = *G. flavomarginatus*; G. a. = *G. agassizii*; and G. m. = *G. morafkai*. The Colorado River separates the two species (G. a. and G. m.) of desert tortoises.

Xerobates; thus, we were back to two genera, *Gopherus* and *Xerobates*, because Agassiz (1857) originally described *G. berlandieri* as *Xerobates berlandieri*, and Brown (1908) selected *G. agassizii* as the type specimen for the genus *Xerobates*. Lamb et al. (1989) and Lamb and Lydeard (1994), using mitochondrial DNA techniques, supported the view of Bramble, but Crumley (1994), after a rather in-depth analysis of morphology and fossils, agreed with Auffenberg. The os transiliens (fig. 1.1), two small bones occurring in the jaw mechanism, tie together the species of *Gopherus*. These bones are uniquely derived characters reported to occur in no other chelonians (Patterson 1971a; Bramble 1974), although someone might wish to recheck members of the genus *Manouria*. The expanded auditory bulla and large sacculus inside that are found in *G. polyphemus* and *G. flavomarginatus* provided the primary morphological evidence for recognizing two genera (Bramble 1982).

Allometry (Huxley and Tessier 1936) is the study of how morphological characteristics of organisms respond to change in another characteristic such as size, that is, differential growth patterns (fig. 1.2). A subtle change in one

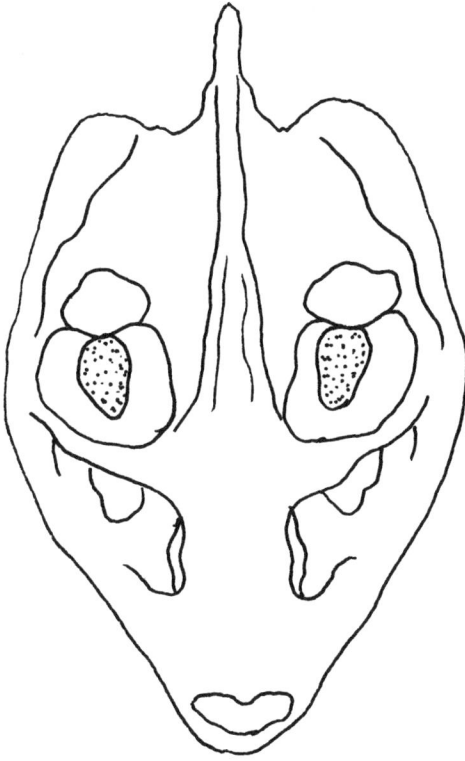

Figure 1.1. Schematic showing the general positions of the os transiliens, tiny bones (stippled) found in members of the genus *Gopherus*. The bones occur more deeply in the jaw than depicted and are somewhat more posterior, making them difficult to illustrate in their proper positions. Image by Roxana Tuff.

Figure 1.2. Allometric growth in the carapace of a female Texas tortoise. Image by Annie Simpson.

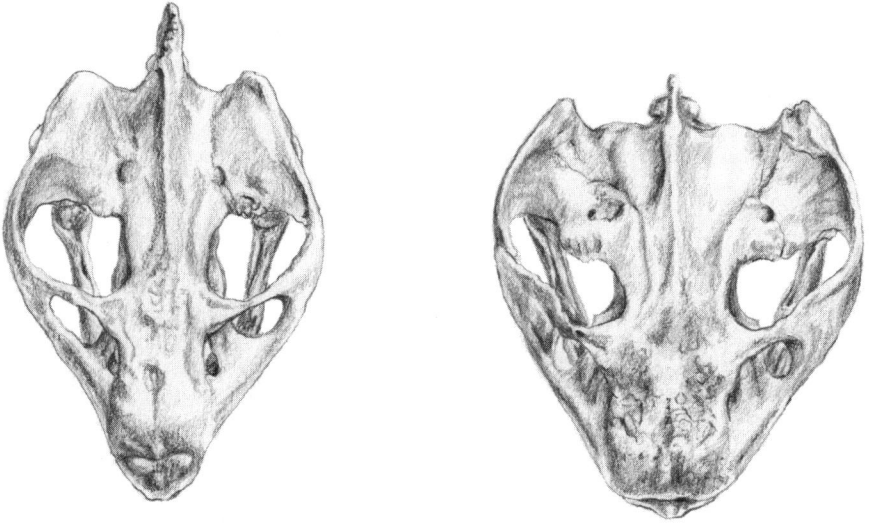

Figure 1.3. Skulls of a Texas (left) and gopher tortoise (right), demonstrating the difference in width. The expansion of the auditory area in the posterior of the skulls probably initiated the corollary changes in other elements. Image by Roxana Tuff.

morphological character might correlate with a significant change in another because growth patterns are interconnected. The skulls of gopher and Bolson tortoises are relatively wider than those of Texas tortoises (fig. 1.3) and are probably reflections of the expansion of the auditory bulla at the posterior of the skull. In order to accommodate the change, the rest of the skull elements would have to alter in concert. The skulls of gopher and Bolson tortoises are thus not simply larger replicas of those of the smaller tortoises; they have acquired proportional differences because of a suite of developmental changes. Similarly, drastic differences in the ontogenetic development of the carapace and plastron bones of male and female Texas tortoises can also be understood as the result of following different allometric paths in response to different intrasexual and intersexual behaviors.

There can also be intragroup hybrids: *G. flavomarginatus* × *G. polyphemus* (Judd and Rose, personal observations of captive animals) and *G. agassizii* × *G. berlandieri* (Woodbury 1952), but no intergroup hybrids have been reported. See Auffenberg and Franz (1978a, 1978b) for supplemental references.

For many years there were rumblings from various researchers familiar with the degree of geographic variation in desert tortoises that more than one species was encapsulated under that taxonomic umbrella. In the mid-1960s, Auffenberg frequently stated that he thought the populations of desert tortoises in Sonora expressed enough morphological differences to be evaluated as a separate species. Berry et al. (2002) suggested that there might be as many as four units equivalent to species represented in the populations in the southern United States and mainland Mexico. Ottley and Velázques Solis (1989) described *Gopherus lepidocephalus* from Baja California Sur, Mexico, based on one living and one deceased tortoise. The recognition, however, was short-lived, and *G. lepidocephalus* was relegated to synonymy under *G. agassizii* (Crumley and Grismer 1994).

Murphy et al. (2011) formalized the description of a new member of the genus, *G. morafkai* (Morafka's desert tortoise) and recommended that the standard name of *G. agassizii* should be Agassiz's desert tortoise. The geographic range of the new species is south and east of the Colorado River (including part of Arizona and the southern tip of Nevada), and in Sonora (including Tiburón Island) and Sinaloa on the west side of the Sierra Madre Occidental (Murphy et al. 2011). Recognition of the new species reduces the known range of *G. agassizii* to about 30% of its former area, and the authors suggested serious implications for the management of this species. Averill-Murray (2011), however, addressed those issues and suggested a more balanced approach that focuses conservation efforts on factors contributing to the declines of both species, including those factors that negatively impact the species' ability to sustain their population levels.

The description of *G. morafkai* (Murphy et al. 2011) came to our attention as we were preparing this book, but the designation was basically foretold by the work of Jennings (1985) and Lamb et al. (1989). One primary objective of these authors was to address the geographic origin of *G. lepidocephalus*, the tortoise found in southern Baja California (Ottley and Velázques Solis 1989) and synonymized with *G. agassizii* (Crumley and Grismer 1994). Even though the section of DNA isolated from *G. lepidocephalus* was short, it was not present in the Sonoran population, confirming its origin in the Mojave population.

Because neither Agassiz nor Morafka actually possesses his namesake species, we suggest that the standard names drop the possessive; the names would then become Agassiz desert tortoise and Morafka desert tortoise.

Householder (1950) reported what he assumed to be courtship and coition between a male and female *G. agassizii*. The male in this coupling was found on a street in Phoenix, Arizona, and had several coats of paint on its carapace. Woodbury (1952), noting that the male tortoise in Householder's published picture looked peculiar, visited Householder and "confirmed" that it was a *G. berlandieri*. Two of the eleven eggs produced by the female were infertile, five were destroyed by dogs, two were destroyed by fly maggots (although we are not enlightened as to how fly larvae were able to penetrate the calcareous eggshell), and two survived. The ratio of two out of nine developing is not out of line with what is expected. However, we must bear in mind that as with humans, in whom morals may be tempered by opportunity, no natural opportunity exists for crossbreeding by either species of *Gopherus*. And, we have no information whether the "hybrids" survived or were themselves fertile. In addition, because females store sperm from previous matings, there is no way to confirm that the attendant male was guilty as charged.

There is thus no credible morphological evidence that *G. agassizii* and *G. berlandieri* hybridize. Certainly the report by Woodbury (1952) should not be accepted as evidence. Edwards et al. (2010) presented DNA evidence that captive tortoises should be suspected of possible hybridization. They found that of 180 captive tortoises from Arizona, more than 40% originated from the Mojave population or were crosses of the Sonoran and Mojave populations. Of the captives in Phoenix, 8.8% were hybrids of desert and Texas tortoises. We have seen photographic evidence of a male box turtle (*Terrapene ornata*) *in copuli* with a female Texas tortoise. Assuming that eggs and viable young were produced, no one would assume that the young were intergeneric hybrids, and that brings us to Boxtor.

Boxtor is a turtle that was assumed to be the product of the female Texas tortoise and the male box turtle mentioned above. It was found and maintained by the Bob and Jean Jones family in San Antonio. Its shell is thickened, the bridge connecting the plastron and carapace is abnormally thick and high, and its hind feet are rounded in outline. Its plastron is also exceptionally thick and has an unnaturally stiff hinge. Observations of multiple copulations between the box turtle and the Texas tortoise, along with Boxtor's abnormal morphology, seemed to be compelling evidence that Boxtor was an intergeneric hybrid, no matter how improbable. Herein lies a warning, because this situation clearly elucidates the difference between a sophistic belief and science: DNA evidence, alas, confirmed that Boxtor was 100% box turtle (M. Forstner, pers. comm.). In addition, photographic evidence confirms that the box turtle's love interest was a male Texas tortoise!

The distinctness of a genus is inversely related to the number of species in that genus, and many genera have only one species (they are monotypic). But, the definition of a genus has changed over the years; it used to mean a taxonomic group of species that are more closely related to each other than they are to other such groups, whereas now it means a group of species that includes the most recent common ancestor and all its descendants (a monophyletic group). To say a group is monophyletic when there are incongruent data sets that may involve morphology, behavior, physiology, paleontology, and genetics (not to mention taxonomy by authority) may again lead us down Ambiguity Lane. But at this juncture, we appear to have settled on the single genus *Gopherus*, which includes two subgenera, one of which contains two recognized species, and the other three—or have we? A review of the genetic literature (Lamb et al. 1989; Morafka et al. 1994) reveals little evidence today supporting the view that the Texas tortoise and Agassiz desert tortoise should be maintained as separate species. Murphy et al. (2011) did not compare data sets that included Texas tortoises. Had they done so, the question would have arisen of whether the genetic differences reported between populations of desert tortoises that led to the

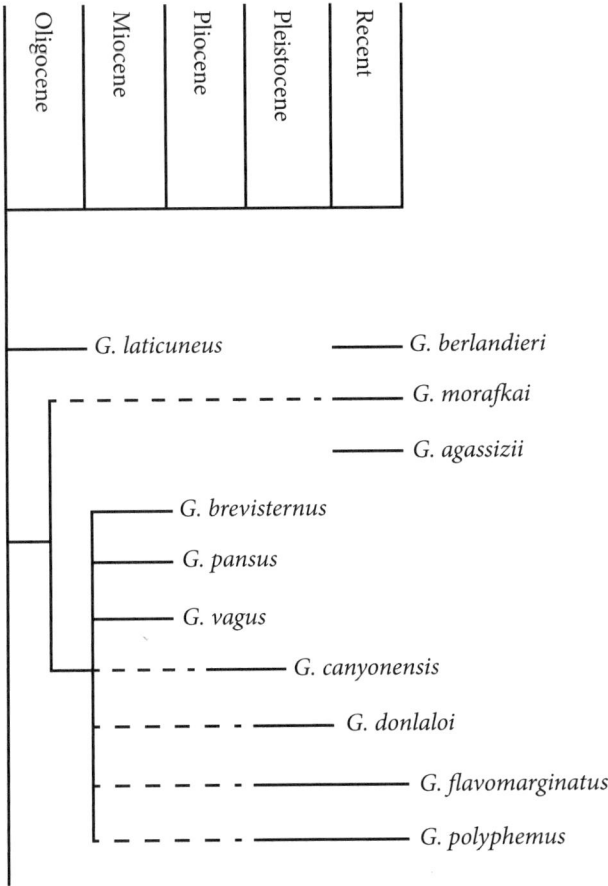

Figure 1.4. Depiction of the phylogenetic relationships of the five extant currently recognized species of the genus *Gopherus*, modified after Reynoso and Montellano-Ballesteros (2004).

recognition of a new species are greater, equal to, or less than those that exist between the Agassiz desert tortoise and the Texas tortoise. If they are as great or greater, then the current specific status should be maintained; but if they are less, then, as onerous as it might be to us, serious discussions should commence regarding the species status of the two groups.

The immediate ancestor of the genus *Gopherus* is probably the extinct genus *Stylemys*, which is found in deposits of Eocene-Miocene age in North America. Although there is only one reported fossil of *G. berlandieri*, there are numerous recognized fossil *Gopherus* species extending back to the early Oligocene, about 34 mya (fig. 1.4). The large Asian brown tortoise, belonging to the genus *Manouria* (plate 2), is probably the closest living relative of the five species of

Gopherus, a well-established view recently supported by Fritz and Bininda-Emonds (2007).

Reynoso and Montellano-Ballesteros (2004) described a large fossil tortoise, *G. donlaloi*, from northeastern Mexico and reviewed the phylogeny and biogeography of gopher tortoises. There is no doubt that members of this group arose in the central Great Plains of North America in the Miocene, but the association of *Stylemys* and *Manouria* with North America is not clear. Reynoso and Montellano-Ballesteros postulated that members of the group extended south into Arizona and Florida and from northern Texas to Aguascalientes, Mexico, during the Pliocene and Pleistocene. To shed some light on the present disjunct distributions of the gopher tortoise in the southeastern United States and its closest relative, the Bolson tortoise, in the Bolsón de Mapimí, a basin in Mexico, Reynoso and Montellano-Ballesteros suggested that the invasion of the Texas tortoise into the realm of the giant gopher tortoise caused extinctions of geographically intermediate tortoises. So, in this hypothesis the smallest of the species, a non–tunnel maker, outcompeted the giant bulldozer tortoises as it invaded their range. We point out that there is only one fossil Texas tortoise known, and that is from Aguascalientes (it was originally designated *G. auffenbergi* but was recognized as *G. berlandieri* by Bramble [1982]), and there is also a possible "robust Texas tortoise" from El Desemboque, Mexico. A "robust Texas tortoise" might be a desert tortoise. Until further fossils are found and placed in proper taxonomic position, its species status cannot be resolved. The Xerobates group probably had its origin in Mexico and has moved northward. It may be that the northern invasion postdates the extinction of the giant members of *Gopherus*. Reynoso and Montellano-Ballesteros did not discuss how glaciation events potentially affected distributions and extinctions. Other hypotheses will be welcomed. They also stated that no fossil of *G. agassizii* is known. We refer the reader to McCord (2002) for informative treatments of fossil *Gopherus*.

If there is uncertainty as to whether there is one genus or two genera, and five species or possibly two species, in this group today, how can we be sure of the taxonomic status of the group and its members 30 million years ago based on fossilized skeletal material? Vertebrate paleontologists and those who prepare fossilized materials for study are remarkably accurate in their taxonomic placements, based on the evidence they have. In addition, they have a limited data set, not complicated by color, behavior, genetics, or geography. In essence, sometimes too many data, like ingredients in a hotdog, stimulate too many sensory modalities and lead to multiple interpretations and the justification of a personal choice.

Texas tortoises are the smallest of the five recognized species and are the most sexually dimorphic. The carapace is olive brown to brown in adults, but the scutes of the young have yellow centers. The gular projection is well developed, much more so in males, and it is forked, which is unique to the Texas tortoise, for in the other species it presents as a single unit. There are two well-

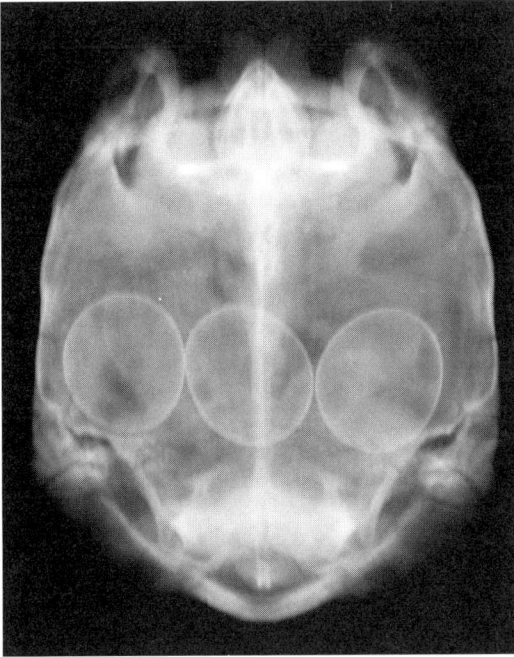

Figure 1.5. X-ray showing shelled eggs inside a Texas tortoise. Note that the three eggs are large and not likely to pass from the tortoise (a condition known as dystocia). X-ray provided by Jeff Jorgenson, DVM.

defined subdentary glands (Smith and Brown 1946; Rose 1970; Rose et al. 1969; Winokur and Legler 1975) protruding from the lower jaw (plate 3) in both sexes, but they are much better developed in males. Hind limbs are elephantine in appearance and forelimbs are spatulate and specialized for digging. The latter are also covered with large scales, and although the tortoise is unable to close its shell, the forelimbs can be drawn inward (plate 4) to form a formidable barrier protecting the head. The well-plated hind feet serve a similar function (plate 5) when withdrawn, protecting the vulnerable inguinal region. Because of the large scales on the forelimbs and the shortness of the hind limbs, the tortoise's gait is stiff and it carries its carapace high when walking. There is a common misconception that the plastron is not hinged, but if that were true, the large, porcelain-hard eggs (fig. 1.5) would not be able to exit the shell (Rose and Judd 1991; Barton 2006).

The posterior plastron is tightly attached but is movable in both sexes (fig. 1.6). There are five toes on each forelimb and four on the hind limbs. Females tend to appear more rounded from the top, while some males are noticeably elongated, and some are saddle shaped. The head is brownish and narrow, but in large males it takes on a more bulbous appearance. Some adults have a dull yellowish patch on the sides of the head, in the temporal region. A rare version occurs wherein the animal is a light tan throughout: these are usually females and are referred to as "blonds." The largest free-ranging male we measured at the Yturria Ranch had a carapace 228 mm (9 in) long and weighed 2.5 kg

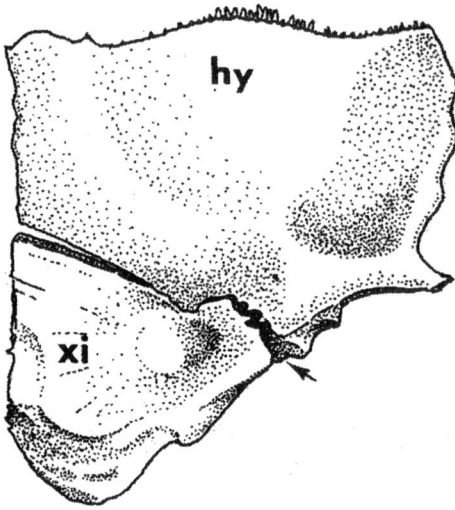

Figure 1.6. The plastron hinge found in female Texas tortoises that facilitates the passing of large eggs; hy = hypoplastron, xi = xiphiplastron; Rose and Judd (1991). Used with permission from the Society for the Study of Amphibians and Reptiles.

(5.5 lb); the largest female had a 210 mm (8.3 in) carapace and weighed 1.65 kg (3.63 lb). The largest male at the Reed Ranch was 238 mm (9.4 in), and two others exceeded the maximum length recorded at the Yturria Ranch. The largest female at the Reed Ranch was 210 mm (8.3 in). Note that these are exceptional animals and that average sizes are considerably smaller. As will be discussed later, inland tortoises are noticeably smaller and less sexually dimorphic than those occurring along the coast. Males have a concave posterior plastron that facilitates coupling with a female (fig. 1.7). In addition, the inguinal region of the plastron near its attachment to the bridge is greatly thickened in males. The animals do not defend themselves by biting but freely express cloacal contents when handled, especially if placed upside down. Although Ernst and Barbour (1972) stated that this behavior is rare, we affirm unequivocally that if you pick up an adult in the wild and turn it on its carapace, you will be plastered 100% of the time. Perhaps it is important to state that the force of these ejecta is directed at a 30° angle to the long axis of the tortoise, so that if you are holding an animal chest high with the plastron facing you, it will be your face that is anointed. If the tortoise has been eating prickly pear tunas, the experience is enhanced by the purplish color and strong odor of the incoming glob of uric acid. Whether this behavior thwarts a wily predator is unknown, but it probably deters a few snowbird tourists who, not knowing the consequences of their action, pick tortoises up along a road.

In a study on the genetic differentiation between northern and southern populations of Texas tortoises, 127 Texas tortoises from Texas were genotyped at 11 microsatellite loci (Fujii and Forstner 2010). Statistical analyses indicated a weak north-south differentiation, with a boundary in southern Duval County. A test for evolution with distance suggested that there had been recent gene flow

R. Tuff

Figure 1.7. Shells of a male (above) and female (below) Texas tortoise showing the high degree of sexual dimorphism in skeletal structure. The sexual differences far exceed those in any other members of the genus and probably exceed those of all other living tortoises. Image by Roxana Tuff.

between the two areas, but the situation was made more complex by human translocation of tortoises. A primary study goal was to test whether the Nueces River basin had been a barrier to dispersal: that hypothesis was not supported. In summary, there is weak genetic differentiation between northern and southern populations of Texas tortoises, at least in Texas.

Chromosomal studies have proved unhelpful in elucidating intergeneric taxonomic allocations or geographic variation in population characteristics of Texas tortoises. The chromosomal complement (Killebrew and McKown 1978) is the same, $2n = 52$, for all species of *Gopherus* examined.

2

Geographic Range and Habitat

The insufferable arrogance of human beings to think that Nature was
made solely for their benefit, as if it was conceivable that the sun had been
set afire merely to ripen men's apples and head their cabbages
—Savinien de Cyrano de Bergerac,
États et empires de la lune, 1656

This chapter will discuss the geographic range of the Texas tortoise and describe the climate and weather, topography, soils, vegetation, seasonal changes, and disturbances in the tortoise's habitat. Understanding the tortoise's environment will provide a foundation for understanding other aspects of its life.

Geographic Range

Comparison of the geographic range of the Texas tortoise with the extent of the Tamaulipan Biotic Province (map 3) shows that the two correspond closely. Indeed, the Texas tortoise may be the most representative animal species of the Tamaulipan Biotic Province (Judd 2002). Dice (1943) limited the northern edge of the Tamaulipan Biotic Province to the southern tip of Texas, but Blair (1950) extended it northward to the Balcones Fault in the north and the line separating pedocal from pedalfer soils in the northeast. This northeastern boundary, near the San Antonio River, also marks where vegetation changes from thorn shrub to prairie and alternating oak and hickory communities (Blair 1950). The northern distribution of the Texas tortoise also appears to be limited by the change from thorn shrub to oak grassland and oak-juniper grassland; that is, the primary habitat of the tortoise is thorn-shrub communities. To the northwest, in Texas, the range appears to be limited by the change from thorn shrubland to desert grassland vegetation.

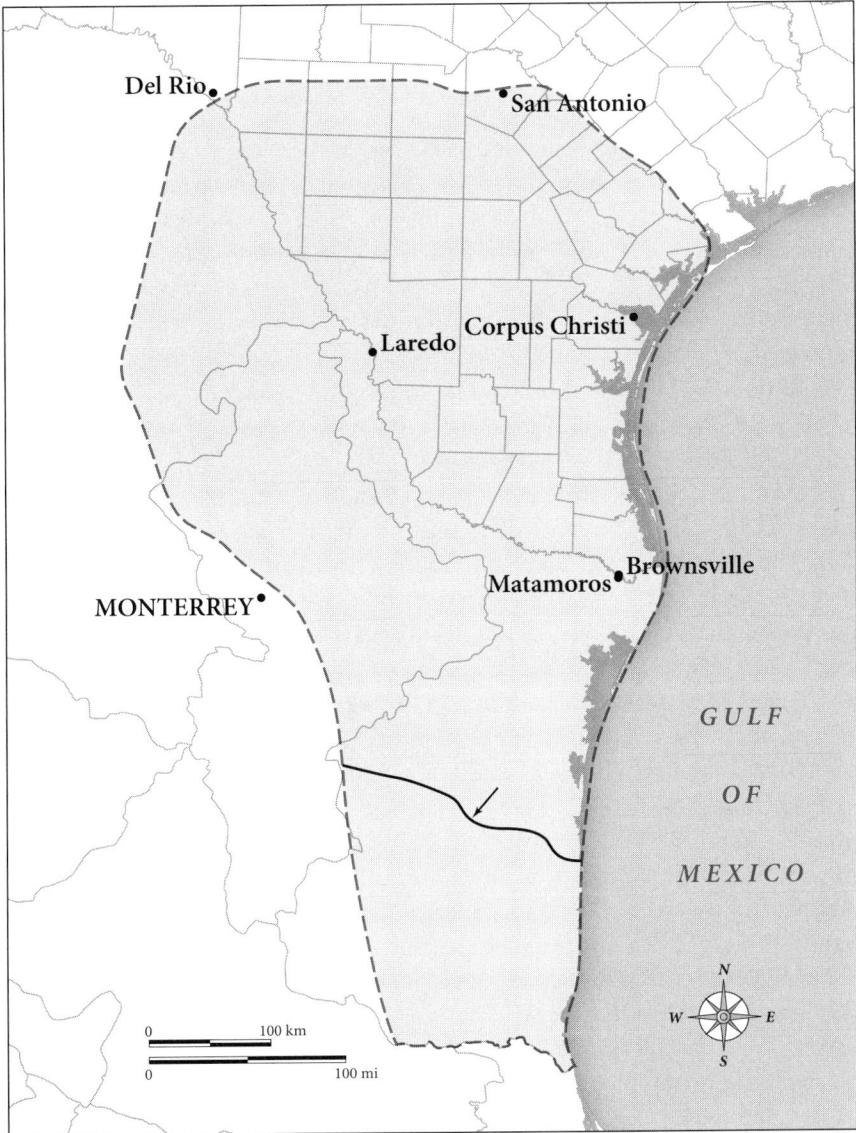

Map 3. General range of the Texas tortoise and the area of the Tamaulipan Biotic Province, showing that they are congruent in the northern 85 percent of the range of the tortoise. The solid line indicated by the arrow is the southern border of the province as depicted by Alvarez (1963). Image by Andy G. Grubb.

Dixon (2000) showed tortoise locations in eight Texas counties north of the northern boundary of the Tamaulipan Biotic Province, but these locations bear question marks and probably represent released individuals. We are unaware of any breeding populations in these counties. Because of the great number of tortoises that are picked up on roads and taken home by travelers, finding released individuals far from the native geographic range is common. The most northern sustained population known to us is also the most western and occurs around Del Rio, Val Verde County. Tanzer et al. (1966) reported a single specimen taken 45 miles (72.4 km) north of Del Rio. Reports of populations in Val Verde County away from the Rio Grande and in Terrell County have not been confirmed. One dead individual was observed near Sonora, Sutton County, and another was recently found in Kimble County (J. Dixon, pers. comm.). Efforts to determine the status of Texas tortoises in Uvalde and Medina Counties (Mittleman and Brown 1947) would be useful.

The Gulf of Mexico forms the eastern boundary of the Texas tortoise's geographic range. The occurrence of tortoises on barrier islands was reported for South Padre Island, Texas (Rose and Judd 1982) and for the Tamaulipan Barrier Island (Selander et al. 1962). Occurrences on these barrier islands may reflect transported individuals. We are confident after working more than 35 years on South Padre Island that there is no breeding population of Texas tortoises on the island. Likewise, we are unaware of any breeding populations in the central and northern portions of Padre Island or on the more northern Mustang, San Jose, or Matagorda barrier islands. Matagorda County (Dixon 2000) marks the easternmost occurrence on the Texas mainland. Strecker (1915) reported a specimen in the Baylor University Museum collections from Jefferson County, which was considered erroneous by Gunter (1945), who stated that Texas tortoises do not occur naturally in Jefferson County or any other part of East Texas. We agree with Gunter's assessment. As an aside, Brown (1950) noted that a specimen of *G. polyphemus* was reported from Jefferson County, "but the specimen upon which the identification was based is not now extant." He then noted that the species should be expected in extreme eastern Texas. We surmise that because phenotypic relationships between Texas and gopher tortoises were not well defined in 1950, there is a high probability that the specimen was a displaced Texas tortoise. The accepted western limit of the range of *G. polyphemus* is extreme eastern Louisiana.

Viosca (1927) reported a Texas tortoise from Avery Island, Louisiana, that was captured in a trap. Avery Island is one of five salt dome islands rising above the Louisiana Gulf Coast and is surrounded by swamps. The sex of the animal was not given, and it was placed in the now defunct Louisiana State Museum in New Orleans. The collection was then sent to the Louisiana State Museum at Louisiana State University, where a tortoise generally matching the description (catalogue no. 457) was identified (by Jeff Mundy) as a gopher tortoise. Whether it is the specimen from Avery Island may never be known, but we agree with Viosca that the tortoise they reported was released at the site.

The western border of the tortoise's geographic range extends southward from Del Rio, in Val Verde County, Texas, and Ciudad Acuña, in Coahuila, Mexico, to slightly west of Nuevo Rosita and Monclova, in Coahuila. Tortoises were found at elevations as high as 884 m (2900 ft) in the foothills of the Sierra Madre Oriental (Rose and Judd 1982, 1989). Vegetation change associated with altitude may set the western limits of the geographic range; that is, tortoise distribution may extend westward (and upward) into the Sierra only as far as the thorn-shrub communities extend. From Monclova, the boundary of the geographic range turns southeastward to just east of Monterrey, Nuevo León, and from there south-southeast to Ciudad Victoria, Tamaulipas. From Victoria the range extends southward into northernmost San Luis Potosí, and then turns eastward to the Gulf Coast. It is important to note that the geographic range does not extend into northern Veracruz. This is likely because marshes around and to the south of Tampico in northern Veracruz constitute inappropriate tortoise habitat. We urge caution in interpreting the southeastern extension of the range in Mexico because we know of no definitive studies on the subject, specimens are few, and notes of direct observations seem to be housed in various scattered field notes of biologists who are no longer with us. Legler and Vogt (2013) reviewed this species in their treatment of the turtles of Mexico.

Climate and Weather

The climate of the Tamaulipan Biotic Province in Texas is semiarid and megathermal (Blair 1950). This means that temperatures are sufficiently high to support plant growth throughout the year, but that at times there is not enough moisture to permit growth.

Southeasterly winds from the Gulf of Mexico prevail in every month except December, when a northwest wind is common. Average relative humidity at 6:00 A.M. at Brownsville, Texas (about midway through the province on a north-south axis), ranges from 85% in December to 90% in July, with an annual average of 88%. Dense fogs are common at night from December through March. This high relative humidity causes fairly heavy dew even during drought years and might be an important free-water source for tortoises. High relative humidity no doubt reduces evaporative water loss for tortoises and may have an important impact on their distribution.

There are two rainfall patterns in the province. First, at any given latitude, precipitation is highest near the Gulf of Mexico and decreases westward. For example, in the northeastern corner of the province, total annual precipitation is 91 cm (35.8 in), and in the northwestern corner it is 51 cm (20.1 in). Second, on a north-south axis, precipitation decreases slightly from the northern boundary to about midway through the province and then increases from the midway point to the south. For example, at Corpus Christi, near the northern boundary, total annual precipitation is 70.6 cm (27.8 in); at Brownsville, near the halfway point, it is 68.2 cm (26.9 in), and at Soto la Marina, in the southern part of the province, it is 74.8 cm (29.4 in) (Tunnell 2002). At Brownsville, about one-third of the total annual rainfall is received in September and October and two-thirds falls from May through October. This division corresponds with the "rainy season" that prevails in the tropical region of eastern and central Mexico (Davis 1942).

Most precipitation comes from thunderstorms, and often a single thunderstorm will account for an entire monthly rainfall. Flood-producing rains can occur in any season but are most frequent in September and October. Heavy rain is often associated with tropical storms and hurricanes.

Rainfall in the province may best be considered irregular and unpredictable at the time scale of seasons, years, and decades. For example, total annual rainfall at Corpus Christi since 1900 has ranged from 13.7 cm (5.4 in) in 1917 to 127.6 cm (50.3 in) in 1992 (Tunnell 2002). A drought in the 1950s lasted 80 months. The long-term average for evaporation from a free-water surface is 147 cm (57.9 in) annually, which is double the average annual precipitation of 58.2 cm (26.9 in) at Brownsville, Texas.

The range of the Texas tortoise (map 3) includes south temperate, subtropical, and tropical climates. The tropical portion occurs from the southern border of the range to the Tropic of Cancer, which lies about 120 km (75 mi) to the north of the southern border of the range. The portion of the range from the Tropic of Cancer north to 26°30' N latitude may be considered subtropical, and the portion north of 26°30' N is south temperate. The factors responsible for setting the northern boundary of the range are unknown, but low winter temperatures likely play an important role. The yearly number of frost-free days in the tropics is 365, in the subtropics it ranges from 330 to 365, and in the south temperate region it averages about 300. Texas tortoises are generally not found north of a line where winter freezes are common and frost-free days are fewer than 300.

The geographic range of the Texas tortoise is characterized by long, hot summers and short, mild winters. San Antonio, on the north-central boundary of the range, has an average winter low of 4.3°C (40.0°F) and an average summer high of 34.3°C (94.0°F). At Brownsville, near the midway point, the average winter low is 10.7°C (51.7°F) and the average summer high is 33.7°C (92.7°F). At Tampico, Mexico, in the tropical portion of the range, the average winter low is 14.0°C (57.3°F) and the average summer high is 31.3°C (88.7°F).

Freezes occur occasionally in the south temperate part of the geographic range, but rarely in the subtropical portion. Two major freezes did occur around Brownsville, Texas, in December of 1983 and 1989. From December 24 to 25, 1983, 55 consecutive hours below freezing were reported at Brownsville, and a minimum temperature of -6.7°C (19.9°F) was sustained for six hours (Lonard and Judd 1985). During the massive cold wave of December 22–24, 1989, two freezing episodes were reported at Brownsville; one with 33.7 consecutive hours below freezing and the second with 16.7 hours below freezing (Lonard and Judd 1991). A temperature at or below -8.4°C (16.9°F) was recorded for 11.7 hours. These two freezes caused major fish kills in the Laguna Madre (Martin and McEachron 1996). If Texas tortoises do not find protection during freezes, the dorsal portion of the carapace to which the vertebrae housing the spinal cord are attached freezes, paralyzing the hind limbs. It is thus probable that such freezes produce mortality in tortoises, but data are lacking on this important aspect of tortoise survival.

Topography

The Tamaulipan Biotic Province (and the geographic range of the Texas tortoise) is confined to the Gulf Coastal Plain of Texas and Mexico, but it also includes numerous rolling hills, some limestone ridges, and a few mountains. The latter include the Anacacho Mountains in Texas, the Sierra de Picachos and Sierra de Papagayos in Nuevo León, Mexico, and the Sierra de San Carlos and Sierra de Tamaulipas in Tamaulipas, Mexico. The tallest of these are the Sierra de San Carlos, which rise to an elevation of about 1534 m (5029 ft). Tortoises occur at lower elevations, up to about 300 m (984 ft), on mountains where thorn-shrub communities exist. The Anacacho Mountains (Kinney County) are intriguing because of their faunal and floristic disjuncts, such as the coatimundi (*Nasua narica*) and orchid tree (*Bauhinia lunarioides*), which are more prevalent in Mexico.

Soils

Soils in the Texas portion of the tortoise's geographic range are mostly mollisols, ultisols, aridisols, and alfisols. Soils in the Mexico portion of the range are aridisols, mollisols, and vertisols. Mollisols are soils of semihumid areas that are prone to calcification. They are rich in bases and their surface horizons are dark brown to black, with a soft consistency. Ultisols occur in humid, warm climates and have a low base content. They are formed by laterization and are intensely leached. Aridisols develop in very dry environments and are low in organic matter and high in base content. They are prone to salinization. Alfisols exhibit translocation of clay elements and well-developed horizons. The humus layer is shallow. Vertisols are dark clay soils that show significant expansion and contraction due to wetting and drying. The barrier islands of Texas and Mexico and a large area on the South Texas mainland have soils composed of either stabilized or shifting sand.

Clearly, tortoises occur on a variety of soils. Texas tortoises make shallow burrows called pallets (plate 6) that are not much deeper than the length of a medium-sized male, about 180 mm (7.1 in). Ease of burrowing may determine the number of burrows in an area and may in turn affect tortoise survival. Tortoise density might be greater where soils are most conducive to burrowing, or where other animal burrows are common.

Biomes and Biotic Provinces

Identifying, describing, and mapping the geographic extent of biomes and biotic provinces constitute methods of organizing the biotic diversity of a continent or area. Dice (1943) introduced the use of biotic provinces to organize biotic diversity in North America. This method focuses on the ranges and centers of distribution of plant and animal species in physiographic regions. Dice defined a biotic province as a considerable and continuous geographic area. A biotic province thus never occurs as discontinuous fragments but can include more than one major ecological association. Freshwater communities, but not marine communities, are considered parts of biotic provinces.

Unlike biotic provinces, which are continuous, biomes are often discontinuous or fragmented. For example, coniferous forest biomes occur on higher mountains in the eastern and western United States but not in the central region where prairie prevails. The recognition of biomes is based primarily on the dominant plants of climax communities but also takes into consideration the animal species present. Biome classification relies primarily on the relative contribution of three plant life forms: trees, shrubs, and grasses. There is no general agreement as to the number of biomes present on a continent such as North America, but most workers recognize between 6 and 11. The geographic range of the Texas tortoise is included in the temperate grassland, desert, and evergreen broadleaf forest biomes (Pianka 2000; Benton and Werner 1974; Krohne 2001). Whittaker (1975) illustrates the position of vegetative communities on a graph of annual precipitation versus average temperature, and this graph is most accurate in placing the tortoise's range in a thorn-forest savanna or thorn-shrub community.

Vegetative Communities

When Dice (1943) first identified the Tamaulipan Biotic Province, his brief description stated that it was characterized by a dense growth of thorny shrubs, small trees, and numerous cacti. True forests were thought to be absent because of high summer temperatures, which induce high evaporation rates; and the presence of limestone substrates, which do not hold moisture well.

Gould (1969) recognized two major vegetation zones in the Texas portion of the Tamaulipan Biotic Province: the Gulf Prairies and Marshes and the South Texas Plains. The distinction between the two zones is based largely on whether trees and shrubs are present. The Gulf Prairies and Marshes zone is primarily grassland, whereas the South Texas Plains zone is thought to have originally supported savanna grassland. Johnston (1963), in his discussion of the past and present grasslands of South Texas and northeastern Mexico, objected to allegations that woody vegetation had encroached on grassland in South Texas and that brush species had invaded from Mexico. He maintained that brush species already occupied their present ranges when the first collector (Jean Louis Berlandier) visited the area in 1828. Johnston also emphasized that many grasslands were infested with mesquite long ago, but that the mesquite growth was stunted due to recurrent fires. He argued that the rapid takeover of grassland by mesquite and other brush species was a result of fire suppression, which permitted an increase in the stature of the aerial parts of plants and an increase in stand density, rather than invasion by brush into previously brushless areas.

McMahan et al. (1984) followed Gould in recognizing the Gulf Prairies and Marshes and the South Texas Plains as the two major ecological areas in South Texas, but within these two areas they identified 12 "vegetation types" (excluding crops). Diamond et al. (1987) retained the same two major ecological areas identified by Gould and McMahan but added a third, much smaller, area, the Coastal Sand Plain, which occurs between the Gulf Prairies and Marshes and the South Texas Plains. The Coastal Sand Plain has evergreen woodland vegetation, and coastal live oak (*Quercus virginiana*) and seacoast bluestem (*Schizachyrium littorale*) are the dominant species.

Vegetation maps can be misleading with respect to the presence or absence of vertebrates such as Texas tortoises. For example, McMahan et al. (1984) mapped the area around Port Isabel and Laguna Vista, Texas, as "Marsh," which one might think would be inappropriate habitat for tortoises. Indeed, there are extensive marshes of gulf cordgrass (*Spartina spartinae*) present that constitute inappropriate habitat for tortoises, but *lomas* (clay dunes), with typical upland thorn-shrub vegetation, are embedded within the mosaic of marshes and salt flats. Indeed, we studied the population ecology of tortoises on lomas on the Yturria and Reed Ranches in this area. Both Auffenberg and Weaver (1969) and Rose and Judd (1982) suggested that gulf cordgrass marshes might serve to restrict the gene flow of tortoises between lomas.

McLendon (1991) classified the vegetation of South Texas but omitted coastal wetlands and the Lower Rio Grande Valley. He divided vegetation into grassland, woodland, and scrubland and recognized 29 communities within 10 associations. No one has systematically surveyed each of the communities identified by McLendon (1991), Diamond et al. (1987), and McMahan et al. (1984) for Texas tortoises, but we have seen tortoises in most of these communities and suspect that they occur in all of them.

Thorn-shrub–grassland communities (plate 7) are widespread in both the Texas and Mexico portions of the tortoise's geographic range. They are usually stratified into an overstory of species such as mesquite (*Prosopis glandulosa*), huisache (*Acacia minuata*), or Texas ebony (*Chloroleucon ebano*), with a dense understory that includes Texas silverleaf (*Leucophyllum frutescens*), coyotillo (*Karwinskia humboldtiana*), granjeno (*Celtis pallida*), guajillo (*Acacia berlandieri*), guayacan (*Guaiacum angustifolium*), Texas lantana (*Lantana urticoides*), leatherstem (*Jatropha dioica*), lime prickly ash (*Zanthoxylum fagara*), lotebush (*Ziziphus obtusifolia*), narrowleaf forestiera (*Forestiera angustifolia*), border paloverde (*Parkinsonia texana* var. *macra*), prickly pear (*Opuntia engelmannii*), anacahuita (*Cordia boissieri*), and blackbrush (*Acacia rigidula*). On gravelly soils where drainage is good, in addition to prickly pear, prominent species include pencil cactus (*Opuntia leptocaulis*), pitaya (*Echinocereus enneacanthus*), lace echinocereus (*Echinocereus reichenbachii*), hedgehog cactus (*Ferocactus setispinus*), and pichilanga (*Mammillaria heyderi* var. *hemisphaerica*) (plate 8).

Shrubs are usually larger near the banks where water stands after rains in areas transected by arroyos. Huisache and retama (*Parkinsonia aculeata*) are often abundant in these places (plate 9). In areas of deep sand, grasses are more abundant and vegetation is usually a savanna type, with large mesquites or oaks scattered in clumps (mottes) or as single individuals (plate 10). Seacoast bluestem and Pan American balsamscale (*Elionurus tripsacoides*) are dominant grass species (McLendon 1991).

Texas tortoises are often found in association with prickly pear. Indeed, Rose and Judd (1975) reported that 62% of the tortoises observed on a study grid were either under or in the immediate vicinity of prickly pear. The cladophylls and fruit of prickly pear provide tortoises with food and water, and cladophylls provide cover and protection. The symbiotic relationship between Texas tortoises and prickly pear will be covered in detail in chapter 3.

Seasonal Changes in Communities

Many trees and shrubs in the south temperate and subtropical portions of the tortoise's geographic range drop their leaves in winter, but some species are evergreen, so shade and concealment are usually available, even in winter. Most trees and shrubs also shed many of their leaves in summer in response to low water availability. This decreases available shade at a time of year when midday exposure to full sunlight for an hour or less can be lethal to a tortoise.

Summer thunderstorms with many lightning strikes can be a source of fire initiation. Tortoise mortality occurs as a consequence of fires, so there is no question that summer drought and drying of grasses can pose a serious danger to tortoises. Prolonged droughts, such as those that lasted for a decade or more in the 1950s and from the 1990s into the early twenty-first century in South Texas and northeastern Mexico, might be times of considerable tortoise mortality because of low free-water availability, low water content in food, and fire. Decreased vegetative cover in both winter and summer may allow tortoises to be more easily detected and to become easier prey for predators. This exposure is important primarily for young tortoises.

Spring rains typically result in the flowering of many herbaceous and woody angiosperm species, but spring rains are less reliable than those occurring in fall. Communities are most lush in the fall, the time of greatest cover and food availability. This is also the time of year when tortoise eggs hatch. So, young are produced at a time when food, water, and concealing cover are high.

Nonseasonal Changes and Perturbations

Droughts, freezes, fires, floods, and hurricanes are the principal natural perturbations that affect tortoise communities. Droughts may last for months or years. At this writing, southernmost Texas has been in a drought for over a decade. Droughts lower water availability for tortoises though a decrease in surface water and water in food, as well as atmospheric humidity, which increases evaporative water loss. Freezes kill tortoises and the vegetation that provides food, water, and concealment.

Fires kill tortoises by sweeping through an area and reducing concealing vegetation (plate 11). The resulting new vegetation, however, may be more palatable and have a higher water content than the mature or senescent vegetation removed by fire. Floods were once common in riparian communities, but these communities are not major tortoise habitat. Dams and other flood control projects (Jarhrsdoerfer and Leslie 1988; Judd and Lonard 2004) have stopped the annual flooding of rivers such as the Rio Grande. Flooding is now a consequence of rainfall over an area rather than rainfall farther up the watershed or moisture from snowmelt. Because of dams and irrigation, river flow may cease and not reach the Gulf of Mexico. Hurricanes affect primarily locations along the Gulf of Mexico and can have direct effects on barrier islands, where few if any tortoises are present. On average a hurricane strikes the Texas coast once every six years (Herbert et al. 2005). Tornadoes associated with hurricanes may affect inland locations. These storms may produce tree blowdown but typically do not cover broad areas. Thus, the effects of wind on tortoise populations are likely minimal. Flooding of tortoise habitat as a consequence of the passage of a hurricane may be a more serious concern, but there are no data on Texas tortoise mortality as a consequence of flooding.

Human perturbations include the clearing of land for agriculture and urbanization, roadway and drainage canal construction, and aerial spraying of pesticides and defoliants. Land clearing has been a major human impact to tortoise populations. Tremblay et al. (2005) documented that 91% of the native woodland that had been present in Cameron County, Texas, in the 1930s had been lost. Unsubstantiated estimates of native woodland loss elsewhere in the Lower Rio Grande Valley range from 95% to 99% (Tremblay et al. 2005). Roads are death traps for tortoises. Unfortunately, there are no quantified data on the number killed annually on roadways, but it must be high. Drainage canals restrict the movement of tortoises and may prevent dispersal from disturbed to undisturbed areas or from high-density to low-density areas. Aerial application of pesticides and defoliants is a common practice throughout the range of the Texas tortoise. No one has reported the effects of drift from these applications onto tortoises and their foods. Thus, we do not know how these chemicals affect tortoise health, reproduction, or mortality.

The Tamaulipan Biotic Province, from a human perspective, is a harsh environment that gives as little of its resources as possible to those organisms adapted to live there, and it takes no prisoners. Seasonal and long-term droughts, unrelentingly high temperatures, and a host of usual and unusual perturbations provide environmental challenges that many organisms, including plants, cannot conquer at all stages of their life history. This environment inspired an anonymous poet to write "Hell in Texas," which likens Texas along the Rio Grande to a proprietary colony of Beelzebub:

> He began to put thorns on all the trees,
> And he mixed the sand with millions of fleas,
> He scattered tarantulas along all the roads,
> Put thorns on the cacti and horns on the toads;
> He lengthened the horns of the Texas steers,
> And put an addition on jack rabbit ears.
> He put devils in the bronco steed,
> And poisoned the feet of the centipede.
> The rattlesnake bites you, the scorpion stings,
> The mosquito delights you by buzzing its wings.
> The sand burs prevail, so do the ants,
> And those that sit down need half soles on their pants.
> And all would be mavericks unless they bore,
> The marks of scratches and bites by the score.
> The heat in the summer is a hundred and ten,
> Too hot for the devil and too hot for men.

With this view in mind, we heed the words of Joseph Wood Krutch, who pointed out in *The Desert Year* (1951) that these harsh lands are paradise to those organisms adapted to them, and "only to those who come from somewhere else is there anything abnormal about the conditions which prevail."

3

Animal Associates and Relationships

Some say it was Bertrand Russell who gave a public lecture on astronomy wherein he described the earth's orbits around the sun and discussed the position of the sun in our galaxy. At the end of the lecture, an older lady near the back of the room stood and admonished, "What you have told us is rubbish. The world is really a flat plate supported on the back of a giant tortoise." He asked confidently, "What is the tortoise standing on?" to which she replied, "You're very clever, young man, very clever, but it is turtles all the way down.

—Cited in Stephen Hawking,
A Brief History of Time

The Texas tortoise, fascinating though it is, does not exist in a vacuum. In this chapter we will examine its relationships with the animals and plants that share its habitat. We will see what the tortoise eats, and what eats it, as well as what challenges it faces from parasites and diseases and how it interacts with its community.

Food and Competitors

As the old saying goes, vertebrate herbivores are born looking for a place to starve to death. This idea reflects the high food consumption of herbivores and the higher energy expenditure needed to process herbaceous material. Texas tortoises are herbivores, which is not to say that they never eat animal protein, because they do. They are basically croppers that feed on small herbs and grasses, but they may also spend considerable time consuming cactus clado-phylls (modified stems commonly called pads) and fruits (tunas) (plate 12). When cactus is present in tortoises' habitat, cladophylls, tunas, and flowers form the primary staple in their diet (Auffenberg and Weaver 1969; Rose and Judd 1982). When tunas are ripe, their purplish color is readily observed in tortoise urine. The cellulose portion of grasses provides bulk to the food bolus but is not itself digested. For whatever reason, Texas tortoises have not developed the microbial relationships that enhance cellulose digestion. A major high-energy component of their diet, therefore, passes nutritionally unused through their system.

Tortoises are frequently observed with large, thick cactus spines penetrating their mouth and neck, but they seem unfazed by this predicament as they go about their daily activities unencumbered by what we perceive as a misfortune. They have been observed eating insects and the fecal material of other animals, notably peccaries (Mares 1971), rabbits (Auffenberg and Weaver 1969), and raccoons (personal observation). Where peccaries are common, tortoises are frequently observed with fecal material from these animals caked around their mouth. Even when provided with high-energy vegetables and fruit in captivity, tortoises consume dead grasses if available, and we surmise that there is some requirement for roughage for efficient movement of food through the intestinal tract.

Schad et al. (1964) reported that bacteria belonging to the genus *Lampropedia* were found in gopher tortoises, but not in desert or Texas tortoises. It is possible that members of this bacterial genus are involved in the breakdown of consumed herbaceous matter. Strangely, the bacteria were found in the intestines of roundworms inhabiting the gut and feces of Greek tortoises (*Testudo graeca*), but there was no evidence that its occurrence was generalized in reptilian herbivores. We mention this because we are aware of no physiological treatments of digestion in the Texas tortoise, and it would certainly be a field worthy of investigation.

Known food items of Texas tortoises include prickly pear (*Opuntia engelmannii*), powderpuff (*Mimosa strigillosa*), leatherstem (*Jatropha dioica*), Mexican evening primrose (*Oenothera speciosa*), sea oxeye daisy (*Borrichia frutescens*), narrow-leaf dayflower (*Commelina erecta*), goldenweed (*Isocoma drummondii*), common wild petunia (*Ruellia nudiflora*), wood sorrel (*Oxalis drummondii*), a variety of grasses including buffalo grass (*Buchloë dactyloides*), and screwbean mesquite (*Prosopis reptans*).

Scalise (2010) collected 51 Texas tortoise fecal samples along a transect that cut across the range of the tortoise. Vegetative analyses were done at five sites where the fecal samples were obtained, and Scalise estimated the percent cover of each plant species. Forb fragments were found in 100% of the fecal samples, prickly pear in 98%, grass in 96%, woody vegetation in 92%, and animal fragments in 57%. He concluded that Texas tortoises foraged selectively ($\chi^2_3 = 875$, $p < 0.001$) and consumed more prickly pear than expected and fewer grasses than expected. Males consumed more prickly pear than females ($\chi^2_4 = 42$, $p < 0.001$) and juveniles consumed more grass and forbs than adults ($\chi^2_4 = 30$, $p < 0.001$). Although it seems a strange conundrum, several studies of tortoise diets confirm that they frequently select nonnative vegetation.

Cactus fruit is a highly nutritious food and has considerable free water (about 85% by weight); however, young cladophylls contain 90–95% water (Rose and Judd 1982). Cladophylls have a high acid content (pH 3.5–4.8), and it is not known how tortoises handle the acid metabolically. Perhaps high blood ammonia levels (Baze and Horne 1970) balance the acid.

Romaine lettuce is the food of choice in captivity, and tortoises will pass over other vegetables and fruit to feed on it first. But it must be said that some tortoises are finicky eaters and will not eat certain common vegetables, while others readily devour them. Free-ranging tortoises do not have this luxury of choice, and they perceive vegetable crops grown within their immediate range as a food resource.

Within its geographic range, the tortoise has few native competitors for food. Cattle and goats may have much the same gustatory habits, and both can eat cactus with spines. However, in ranch country where these mammals coexist with tortoises, cactus is not limiting. In the absence of cactus, with high grazing pressure, competition could be substantial, but we have never observed a situation where grazing ruminants were in sufficient densities to negatively impact tortoises. In fact, although the destruction of thorn-shrub areas to develop pastures is initially harmful to tortoises, the established pastures are beneficial (Auffenberg and Weaver 1969; Rose and Judd 1982).

While we understand that grasses, forbs, and the fruits of various woody plants and cacti are important components of the tortoise's diet, there is little information available concerning the nutritive value of these food items, or whether tortoises select food based on nutritional content, as some aquatic turtles do (Fields et al. 2003). There would be no information on the subject at all if it were not for the economic value of deer and cattle in South Texas (landowners raise cattle for sale and derive income from deer leases, which allow hunters to hunt deer on their property). Everitt and Alaniz (1981) analyzed the nutrient content of 17 species of woody plants and cactus fruit (*Opuntia lindheimeri*, or *O. engelmannii*) from South Texas and reported that 15 species contained more crude protein than cactus fruit (which had 7.06%). Twelve species contained more phosphorus than cactus fruit (which had 0.15%); however, only four species had more calcium (cactus had 2.43%). The highest values for magnesium (0.93%) and potassium (3.41%) were obtained in cactus fruit, where sodium levels were average (0.01%). The phosphorus to calcium ratio was 0.06. Thus, one of the primary seasonally available foods of tortoises and one they readily eat is low in protein but high in some minerals. Calcium is probably not limiting to tortoises in South Texas, but the metabolic interplay between calcium and phosphorus in the soil and vegetation is not known and is in need of detailed study.

In x-raying female tortoises to determine clutch and egg size, we noted that some females contained snail shell fragments in their intestinal tracts (Judd and Rose 1989). Therefore, we examined x-rays of 90 females and 16 males to see if they ate snails. Eighteen (20%) of the females x-rayed in 1986 and 1987 contained snail shells. No males contained snail shells, and no shells were found in females after June 21, 1986, or after June 24, 1987. The absence of snails corresponded with a sharp decrease in the percentage of females with shelled eggs that we saw beginning in July of both years. Thirteen of eighteen females with snail shells contained eggs (72.2%). Snails likely provide calcium for eggshell production and protein for vitellin synthesis, but it is not known if live snails or their remnants are the gustatory targets. We were unable to entice captive tortoises to eat live snails, whereas omnivorous box turtles readily devoured them. We did not find evidence of rocks in the intestinal tracts of any tortoises; however, Diaz-Figueroa and Mitchell (2006) and Barton (2006) showed x-rays of Texas tortoises with rocks in their intestinal tracts. We assume these to be captive tortoises under a veterinarian's care, and Diaz-Figueroa and Mitchell stated that the rocks were ingested accidentally.

Predators

There are a host of potential predators capable of killing and devouring Texas tortoises. Notable among mammals are coyotes (*Canis latrans*), raccoons (*Procyon lotor*), opossums (*Didelphis virginiana*), foxes (*Urocyon cinereoargenteus*), skunks (*Mephitis mephitis*), bobcats (*Lynx rufus*) and feral hogs (*Sus scrofa*). Wood rats (*Neotoma micropus*), with whom tortoises cohabit in cactus patches, are known predators of small tortoises, and adults frequently exhibit telltale chew marks on their carapaces. We suspect that these injuries occur mostly in winter and are the result of the rodents' effort to secure calcium, for they frequently drag bones back to their middens to chew on them. William Weaver (pers. comm.), while studying Texas tortoises, observed wood rats capturing a yellow mud turtle (*Kinosternon flavescens*) at the edge of a pond and dragging it back toward a midden.

While wood rats dine delicately, hogs slurp down most things that walk, fly, crawl, swim, or dig a hole. Culbertson (1907) gave a firsthand account of watching hogs devour box turtles (*Terrapene carolina*) that were gathered in a puddle because of a heat wave. This feral predator long ago passed through the veil of control and roams over Texas as essentially a giant gobbler of our native heritage and resources.

Of course, smaller tortoises are more susceptible to predation, and this vulnerability may explain, in part, the secretive behavior of hatchlings and subadults. Snakes, notably western diamondback rattlesnakes (*Crotalus atrox*) and indigo snakes (*Drymarchon corais*), are known predators of small Texas tortoises, and it is likely that birds of prey will also partake of a juvenile tortoise meal. Gulls along coastal areas conceivably take small tortoises, but their impact would be nil. Crested caracaras (*Caracara cheriway*) have increased in numbers in South Texas in recent years and have the demeanor and machinery to make short work of a small tortoise. Dogs associated with human habitation are a known cause of tortoise injury and death. Some dogs are large enough to puncture the shell, instigating serious internal infection. Their handiwork usually includes the gnawing away of peripherals and the anterior and posterior plastron. Foxes, rats, and raccoons gnaw at the front legs (plate 13) until enough damage is done to get at the tortoise's head. Then, the shell is essentially stripped of its contents. If harried long enough, many aquatic turtles and some tortoises seem to "forget" that their best defense is to remain tightly tucked inside their shell, and they attempt to flee, exposing their limbs and head to their aggressor.

Large mammalian predators that could prey on tortoises, such as wolves and mountain lions, have been seriously reduced in number through overhunting. This reduction in large predators has lifted a restraint on the number of meso-mammalian predators, along with subsidized feeding via pet food left outside in urban areas and the reduction in trapping for pelts. Foxes are probably more efficient at killing and extracting tortoises and other turtles from their shells, but raccoons receive the most press. This may be because they are more common and visible than foxes, which are also masters of deceit. As others have pointed out (Dodd 2001), raccoons learn to kill turtles, and each successful capture increases their efficiency. Progression from killing and eating juvenile turtles to killing and eating adults is to be expected, and once a raccoon attains this level of efficiency, little will help but to eliminate the animal. Thirty-two sick and injured Texas tortoises were maintained in an external enclosure (1/10 of an acre square) known as the Texas Tortoise Ranch (not to be confused with the famous Texas Chicken Ranch) in the Texas Hill Country. Raccoons visited the facility after about a month, killing and devouring several small adult tortoises. Within another month, 32 raccoons and 2 foxes were removed from the site.

Parasites

Remarkably, Texas tortoises are relatively free of parasites. We have never observed a tick attached to a tortoise, but chiggers (*Eutrombicula alfreddugesi*) have been reported (Goff and Judd 1981). A few animals were found parasitized by sarcoptid fly larvae (Neck 1977), and while not pleasant to look at, these larvae are not life threatening, although they may weaken their host. After intense combat, some male tortoises tear the skin at its attachment with the anterior part of the carapace. Usually this dries quickly, but if not, sarcoptids may lay their eggs along the crease. The maggots are small and can keep the wound open and draining for some time, but eventually they drop off and the wound closes. In other cases a small round nodule appears (usually on a hind limb), and on close examination a small central opening can be seen. The nodule is filled with chalky debris and usually has only one maggot inside. It might be tempting to remove the material and maggot, but this might result in breaking the capsule lining, which could induce septicemia and lead to the death of the tortoise. It is best to have the tortoise treated professionally unless it is free ranging, in which case it should be left to its own devices: the interaction is just part of the panoply of life.

On occasion, the end result of an injury followed by fly larvae infestation may be severe structural damage. We assume that the actual damage is due to a bone infection (plate 14), and while it is difficult to contain, it usually runs its course without killing the tortoise. Larger tortoises appear to withstand such infections better than juveniles.

One unpublished study done in the 1960s by Danny Pence and Stanley Casto looked for internal parasites but revealed no flukes or tapeworms in about 20 tortoises, although several small roundworms were observed. These small, immature oxyurid nematodes probably represented accidental infestations, and tortoises may not be a definitive host. However, McAllister et al. (2008) reported that ova and a single adult female nematode in the genus *Alaeuris* passed in the feces of a single Texas tortoise from Webb County. One wonders if the coarse diet of wild tortoises might enhance elimination of alimentary tract parasites; nematode loads in captive tortoises can become substantial.

When we established a study area on the Reed Ranch in Cameron County in 1977, we found several tortoises with small holes in their peripherals, a common marking method. This perplexed us because it appeared that we had established a study area over a previous one we knew nothing about. The landowner assured us that no one had studied tortoises on his property before. He casually remarked that a group from the Centers for Disease Control (CDC) had visited the area several summers before and had taken blood samples of every vertebrate captured in order to look for virus reservoirs, because there had been a severe regional outbreak of equine encephalitis. Eventually we found a tortoise with a numbered CDC tag. On contacting an acquaintance who worked for the CDC, we were informed of the CDC's efforts at the site and were told that Texas tortoises were found to harbor the virus. Bowen (1977) concluded that the "Texas tortoise could serve as an overwintering reservoir for WEE (Western Equine Encephalitis) at any temperature."

Diseases and Injuries

A primary problem facing all species of tortoises is a series of viral and bacterial diseases that induce similar symptoms of breathing distress and nasal drainage. Upper Respiratory Tract Disease (URTD) is a generic name for a group of disease symptoms that affect the upper, and occasionally the lower, respiratory tracts of tortoises. These diseases are transmissible (Jacobson et al. 1991), debilitating, and sometimes lethal. Symptoms in desert tortoises are known to be associated with *Pasteurella testudinis* (Snipes and Biberstein 1982) and *Mycoplasma agassizii* (Jacobson et al. 1991; M. Brown et al. 1994; D. Brown et al. 2002). The herpesvirus should also be suspect (Jacobson et al. 1991; A. Johnson

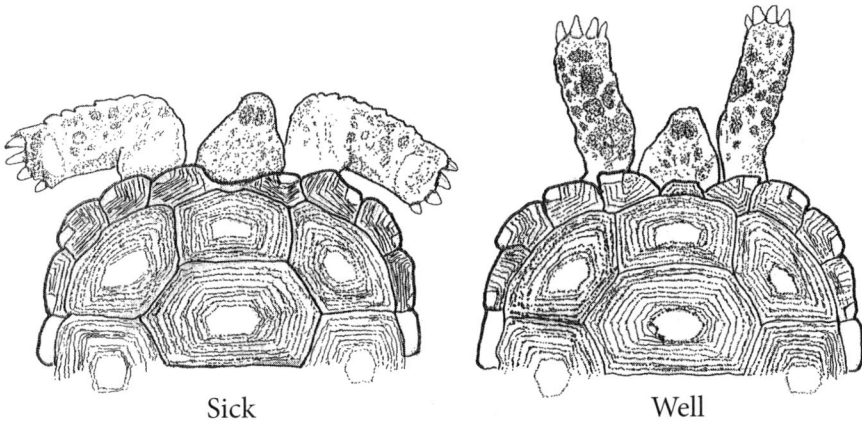

Sick   Well

Figure 3.1. Differences in basking posture of a sick (left) and well (right) Texas tortoise.

et al. 2005), as well as iridoviruses (Westhouse et al. 1996). Iridoviruses of the genus *Ranavirus* have a history of causing recognizable mortality events in several vertebrate groups, including reptiles (A. Johnson et al. 2008; Allender et al. 2011), and are the causative agents leading to high mortality in young frogs, salamanders, and box turtles in Montgomery County, Virginia (Richard Seigel, pers. comm.). These and other disease organisms cause overlapping symptoms, but to ascertain what is going on in Texas tortoises will require thorough necropsies and histopathology. Treatment of this disease conglomerate via antibiotics and vitamins is ineffective.

In 1993, 32 captive Texas tortoises were sent from various sources to Texas State University, where they were maintained in an outside enclosure (Texas Tortoise Ranch). One long-term captive from El Paso and one from Tomball, Texas, were lethargic and had copious nasal drainage several weeks after arrival. Within six months, only 16 tortoises were alive, and all exhibited similar symptoms. The tortoises were moved to a smaller enclosure where they could be treated and observed more readily, and all but seven died. They were allowed to pass the winter in the outside enclosure, and when they emerged in the spring, all exhibited symptoms; two died and five have been symptom free since that time. The actual cause of death is unknown, but a blood sample tested positive for *Mycoplasma*. Necropsies showed that respiratory tracts were clear, with no signs of inflammation or necrosis; the lungs were clear. Individuals steadily declined in vigor and the positive response to antibiotics decreased between each bout. In the final stages, the tortoise splayed its forelegs outward (fig. 3.1) with its head exposed and on the ground; when the body was lifted, the legs hung freely. Decomposition was extremely rapid once death occurred, much more so than if the animal had been killed by another cause. In the vernacular, it was

Figure 3.2. Fungal infection (*Fusarium semitectum*) on a Texas tortoise. Most adult tortoises show evidence of this infection. The fungus catabolizes keratin, the protein component making up scutes and nails, but does not appear to be lethal. Image by John Senter.

resolved into a foul-smelling "soup." This suite of diseases might reflect the adage "Get well or die," but our concern is that symptom-free tortoises released to the wild might serve as carriers. Because tortoise combat and reproductive behaviors involve frontal attacks, pathogen transmission would be assured. Tristan (2009) reported that 39 Texas tortoises from a wild population were seronegative for *Mycoplasma agassizii*, and 12 of 15 (80%) housed in a rehabilitation facility were seropositive. He concluded that Texas tortoises could mount an immunological response to this disease. He suggested, and we agree, that further studies are warranted.

Many tortoises exhibit small whitish spots on the scutes. Known as necrotizing scute disease (Rose et al. 2001), this condition reaches its most extreme manifestation in Texas tortoises (fig. 3.2) A keratin-catabolizing fungus (*Fusarium semitectum*) was identified associated with these lesions and was found to be able to induce the condition. Many pathogens are opportunistic and readily invade exposed lesions, making identification of the causative agent difficult. We suspect that the dry environment inhabited by Texas tortoises limits microbial action that would be more aggressive in moist environments. Necrotizing scute disease does not appear to be life threatening, as is dyskeratosis in desert tortoises (Jacobson et al. 1994). Note that human fingernails and eye corneas are keratin based, and care should be taken to ensure that children or immunocompromised individuals do not handle infected tortoises.

Figure 3.3. Cystic calculus (bladder stone) from a Texas tortoise. The calculus is composed of concentric rings and probably takes years to form. It is obviously too large to pass from the bladder. Image by John Senter.

Bladder stones (cystic calculi) are uncommon in Texas tortoises and may be more of a problem in captive individuals. These stones can be quite large (fig. 3.3) relative to the size of the tortoise and appear to develop over several years (Silverman and Janssen 1996). Certainly, diet and hydration regimes foster the growth of the stones in captive turtles, but it is probable that the extremes of low hydration experienced by Texas tortoises in South Texas and northeastern Mexico induce the condition in free-ranging individuals. However, none of the 77 free-ranging tortoises that we x-rayed in our study of egg production contained dense masses attributable to bladder stones. Texas tortoises form significant amounts of uric acid as a waste product. We posit that dehydration with concomitant high levels of protein in the bladder fluid allows the formation of a primary nucleus that accumulates more material, leaving discernible rings when the stone is cut in half. A calculus 70 mm (3 in) in diameter was recovered from a tortoise shell (with a carapace length of 200 mm [7.9 in]) found in Cameron County, Texas. A male (with carapace 138 mm [5.4 in] long) from Jim Hogg County was maintained in captivity for about 30 days after he became listless and responded weakly to physical stimuli. All appearances indicated that the animal was dehydrated, but efforts to hydrate and force feed him were unsuccessful: he died within one week. A calculus 40 mm (1.6 in) in diameter was observed in the urinary bladder.

Tortoises sustain and survive serious and disfiguring injuries. A prime example was Mashed Morris (plate 15), found in Cameron County. We assumed that his injuries were the result of being struck by a vehicle. Auffenberg and Weaver (1969) commented on tortoise injuries due to National Guard exercises in one of their study areas. In most vertebrates the spinal cord is tightly housed inside the vertebral canal, and injury and swelling of the cord quickly leads to permanent paralysis, for there is little room for expansion. The spinal cords of Texas tortoises are quite small and the vertebral walls are thin, but it is difficult to fathom why such massive injuries do not compromise life; perhaps they demonstrate the tortoise's ability to accommodate spinal cord swelling.

Tortoise injuries from predators and pets are common and are usually manifested in damage from bites to lateral aspects of the shell and gular projections. Carapace puncture occurs even in large tortoises, but surprisingly, even these injuries may not be life threatening.

Wood rats leave their telltale tooth marks on carapaces, and we assume that they do this during the tortoises' winter retreat inside the wood rats' middens. In one group of captive tortoises, Norway rats (*Rattus norvegicus*) gnawed the forelimbs enough to expose the bone, so we assume that small captive tortoises are fair game for rats.

One captive tortoise given to us was in the process of sloughing off its outer skin from along its neck and from all four limbs. The skin surfaces were "weepy" and the tortoise was undoubtedly in deep distress. We eventually surmised that the "burns" were chemical and were the result of fertilizer exposure. Apparently, the grass where the tortoise was allowed to roam had been treated with a dry, slow-acting fertilizer that was high in nitrogen. When the tortoise walked through the dew-moistened grass, it accumulated enough dissolved fertilizer to induce burns. Such an injury could foster microbial invasion and should be treated vigorously.

Failure to thrive is at once an interesting and sad condition noted in hatchlings. The condition is rare in captive animals and manifests as a normal tortoise, though it is generally small and late to hatch, that eats sparingly or not at all, though it may drink. Instead of the shell hardening with age, the tortoise becomes noticeably soft, as bone formation is retarded. Many hours have been expended caring for these animals, usually for naught, as they soon succumb. Necropsy generally reveals gut narrowing, but whether this induces the condition or is induced by not eating is not resolved. Whereas images of these tiny, helpless creatures play on the heartstrings of caregivers, the reality of their fate is that they will get well or die, depending on the roll of their genetic dice, or the bad luck of environmental factors that negatively impact complete development. Because they may be hatched from an otherwise well-developed clutch in captivity, where hatching regimes are tightly controlled, we suspect that genetic factors incompatible with life are the root cause.

Fire-induced injuries are observed and in severe cases scutes and underlying bone are destroyed. If the inner bone is not destroyed, a new shell portion, complete with an amorphous scute cover, is formed (plate 16). These lesions may be small or large, but we have never observed them to the extent that they occur in some box turtles where the total carapace is replaced (Smith 1958; Rose 1986). With the advent of the thousands of acres of buffelgrass (*Pennisetum ciliare*) and guinea grass (*Panicum maximum*) that are now established throughout much of the range of the Texas tortoise, we expect that fire-induced deaths and injuries will increase. Guinea grass has a weak growth habit, but it grows tall and thus intertwines with the supporting branches of thorn shrubs. This places dry thatch well into the plant canopy, forming a secondary fuel source that can enhance the devastating fires that engulf slow-moving organisms.

The following is a partial list of the known diseases and compromised conditions found in Texas tortoises: abscesses, arthritis, constipation, diarrhea, egg retention, eye infections, impacted colon, parasites, pneumonia, renal failure, respiratory disease, septicemia, shell rot, skin infections, stomatitis, and dystocia.

Dystocia is a condition in which the reproductive products are unable to pass from the body, or do so with difficulty. In reptiles this condition is commonly referred to as being *egg bound*. Generally, the female presents with respiratory distress, exhibiting frequent short breaths. An x-ray is necessary at this stage (see fig. 1.5). Because of the boxlike nature of the tortoise's shell, the hard-shelled eggs pose a problem if they are too large, because a basketball cannot pass through a doughnut hole. There are several levels of concern. If the oviducts are not contracting properly, hormone therapy is in order. A combination that works well for Texas tortoises is 5–10 units/kg of oxytocin followed in 10–20 minutes by prostaglandin F2 alpha at 1.0 mg/kg (M. L. Feldman and E. M. Feldman, pers. comm.). If the pelvic aperture is too small, then serious ramifications can develop.

It is possible for the egg of a Texas tortoise to pass through the pelvic girdle but not be able to exit the shell through the gap between the carapace and plastron. Bear in mind that the cloaca of a Texas tortoise is too small to accommodate multiple eggs simultaneously. If a person lubricates a fingertip and gently probes and massages the cloacal opening, the female will usually relax and enable the person to see if an egg is present in the cloaca. Although it seems impossible, a finger can be inserted into the cloaca, and if the egg is maneuvered toward the plastron with the finger on the other side of the egg, it can be crushed. Small eggs are difficult to break in this manner but can be punctured with a probe. Once punctured, they are more easily crushed. By curling the finger in a circular motion, one can sweep the contents outward. The procedure is repeated with the remaining eggs. It has not been possible to confirm instances of egg binding in Texas tortoises in wild populations; the condition is known only in captive animals, where high-energy diets are combined with little physical activity.

Mutualistic Relationships

Mutualism refers to an association between a pair of species that results in benefits to each. Mutualistic associations are said to be facultative if both species can exist without the other, or obligate if neither species can live without the other. Far more attention has been devoted to the study of competition and predation than has been devoted to mutualism. Consequently, mutualistic interactions between reptiles and other organisms might be more widespread than previously thought (Zug et al. 2001). In the course of our studies of Texas tortoises, we observed a facultative mutualistic relationship between tortoises and prickly pear. Closer examination revealed a much more complex interaction among various species, which we call the cactus-tortoise-wood rat-rattlesnake community (fig. 3.4).

That tortoises eat prickly pear is well established (Auffenberg and Weaver 1969; Rose and Judd 1982). Tortoises consume the cladophylls of prickly pears, but the ripe, dark red to purple fruits are preferred. In the process of consuming fruits, tortoises swallow seeds. Rose and Judd (1982) showed that seeds passing through a tortoise's gut had a tenfold higher germination rate (17%) than untreated control seeds (1.7%). Seeds passing out in a fecal pellet may have a selective advantage because the seed coat is scarified and allows the seeds to germinate more quickly when water becomes available, and also because the fecal pellet provides a nitrogen source to enhance early growth. Thus, prickly pear benefits from the association in that it obtains dispersal of its seeds by the tortoise coupled with a higher germination rate and more favorable conditions for early growth after germination. Tortoises benefit from the cactus in that they obtain food, water, concealment, and shade. Indeed, Rose and Judd (1975) reported that 62% of tortoises found on a study grid were either under or in the immediate vicinity of prickly pear.

This relationship is made more complex by the presence of southern plains wood rats (*Neotoma micropus*), which build dens composed of twigs, branches, old cladophylls, and various items of debris in, around, and among clumps of prickly pear cladophylls. Underneath this aboveground midden is a system of underground tunnels. Tortoises may be found pushed up inside a midden or in tunnel openings at any time of the year, but in winter, when tortoises are inactive for lengthy periods, wood rats may expand the midden with additional twigs and branches so that a tortoise is completely covered. Covered tortoises thereby obtain insulation from cold and concealment from predators. In turn, wood rats gnaw on the shells of dead tortoises that they bring back to the midden, as well as on those of live but inactive tortoises in the midden, and presumably obtain calcium from the tortoise shells.

Wood rats obtain both food and water from prickly pear cladophylls and fruits (Braun and Mares 1989). Prickly pear may be the chief source of water for wood rats during periods of drought in South Texas. Whether prickly pear

Figure 3.4. The cactus-tortoise-wood rat-rattlesnake community. This unstudied community is poorly understood. Image by Roxana Tuff.

seeds that pass intact through the gut of wood rats have a higher germination rate than those falling to the ground is unknown, but clearly wood rats benefit from the association, for they obtain protection, materials for den construction, food, and water from prickly pear. Thus, there is certainly a commensal relationship (one organism benefits), and there may be a mutualistic relationship (both organisms benefit) if prickly pears obtain better dispersal of their seeds and a higher germination rate from seeds being scarified by passing through the guts of wood rats. Raun (1966) found a positive correlation between prickly pear presence and wood rat densities.

In South Texas, wood rats are the most common food item of diamondback rattlesnakes (*Crotalus atrox*) (Cottam et al. 1959), yet Pérez et al. (1979) reported these rats to be resistant to the snake's venom. And, like wood rats, diamond-back rattlesnakes also have a commensal relationship with prickly pear. They are commonly found among the cladophylls of a cactus clump. In summer they obtain shade, in winter they get insulation from the cold, and throughout the year they benefit from concealment and protection, especially in winter, when the low temperatures make them especially vulnerable to predators.

The cactus-tortoise-wood rat-rattlesnake community is made more complex by a variety of species that take refuge in wood rat middens. Included among these are least shrews (*Cryptotis parva*), desert shrews (*Notiosorex crawfordi*), and a variety of snakes and lizards. As a cautionary note, although hantavirus in wood rats (Dearing et al. 1998) has not been documented to be prevalent in Texas, we recommend care when disturbing middens. Cone-nose bugs are vectors (Kjos et al. 2009) for *Trypanosoma cruzi*, the agent of Chagas' disease, and are known to use Texas tortoises for blood meals. The association of these bugs, wood rats and their middens, Texas tortoises, and Chagas' disease should be evaluated.

Place and Role in Community

Texas tortoises are primary consumers. Prior to the occupation of their geographic range by Europeans, Texas tortoises likely ranked fifth in importance to bison (*Bos bison*), white-tailed deer (*Odocoileus virginianus*), pronghorn (*Antilocapra americana*), and collared peccaries (*Tayassu tajacu*) in consumption of plant biomass in the community. Adult black-tailed jackrabbits (*Lepus californicus*) weigh 1.5 to 4 kg (3.3 to 8.8 lb) and are therefore as heavy as or heavier than Texas tortoises, but they are not nearly as abundant. Consequently, they are likely less important than tortoises in terms of their biomass and importance as grazers. Our hedge in the previous statement is because jackrabbits eat more than tortoises, so their total consumption of plant biomass over an extended period may be greater.

Predation on Texas tortoises was discussed previously in this chapter, but it is important to note again that predation is primarily on eggs and young. Once tortoises attain a carapace length of about 100 mm (3.9 in), less mortality is attributable to predation.

4

Morphology and Anatomy

When I was young, Statistics was the science of large numbers. Now, it seems to me rapidly to be becoming the science of no numbers at all.
—Oswald George in Roy M. Chiulli,
Quantitative Analysis: An Introduction

Details of the basic body structure and organ systems of the Texas tortoise will be discussed in this chapter, including the skeletal, circulatory, digestive, respiratory, urinary, and reproductive systems.

Skeletal System

The skeletal systems of tortoises are organized into two major components, the exoskeleton and the endoskeleton. The ribs and their bony extensions make up a major portion of the exoskeleton, which when coupled with the presence of a plastron, a feature not observed in other reptiles, presents a unique situation in which demarcation of the two systems is fuzzy at times. A tortoise can be perceived as a bony box with ports for the limbs, neck and head, tail, and penis. Vertebrate skeletal systems are divided into the axial system, which includes the cranium, associated jaw and hyoid, vertebrae, and ribs; and the appendicular system, which includes the pectoral and pelvic girdles and the limbs associated with those girdles.

The shell of a turtle is one of nature's architectural marvels. It is a bony encasement made of approximately 50 pieces in the dorsal domed portion (carapace) and 8 or 9 pieces in the ventral or bottom portion (plastron). Though it would seem that the number of bony elements and their configuration and participation in shell formation should be invariant, such is not the case, as is demonstrated by various innovations in the shells of leatherback sea turtles, soft-shelled turtles, snapping turtles, and pancake and hinged tortoises.

Axial Skeleton

Carapace.   Carapaces (figs. 4.1 and 4.2) of Texas tortoises are formed from 49 bones: along the central long axis there are 7 to 10 unpaired *neurals* that anchor internally to the spinal column vertebrae, an anterior *nuchal*, two *suprapygals*, and a terminal *pygal*. Sixteen *costals* (plate 17) incorporate the ribs and make up the bulk of the dome. Rib remnants are seen internally at their attachments to the neurals and laterally as a point that nestles into a socket in the appropriate peripheral bone. Up to four rib heads may anchor on a conspicuous knob on the eighth costal. The pygal, suprapygals, and last neural do not have costal involvement. Rib heads associated with the first neural are substantial because that area serves as the attachment for the cervical (neck) vertebrae. The second to eighth costals are wedge shaped, with the narrow ends of the second, fourth, and sixth alternating with the wide ends of the third, fifth, and seventh. As viewed externally, the fifth and seventh costals frequently appear not to reach the peripherals on one or both sides, but internally they do.

The eighth costal anchors the pelvic girdle to the carapace, and at this attachment the posterior flared portion of the pelvic bones frequently wears through the bony shell (plate 18), leaving only the thin scute covering to separate the interior of the tortoise from opportunistic pathogens. This condition occurs in adults of both sexes but is prevalent in females, which have thinner bony elements ostensibly because they need their calcium to form eggshells.

Carapace elements are held together by 23 peripheral bones, the pygal, and the nuchal. The third to sixth peripherals form the bridge that connects the plastron and carapace.

Auffenberg (1976) reported that 72% of the Texas tortoises he examined had either abnormal carapace bones or scutes, and the primary abnormalities occurred with the number and arrangement of the suprapygals and whether or not some peripherals were fused. Since there are 49 bones and various ossification centers that have to form a near-hemispheric configuration, we would expect that numerous "mistakes" would be made. Like the work of a good carpenter, these mistakes are covered up with trim, in this case the hornlike, keratin-based scutes. These scutes occur in a regular and programmed formation (figs. 4.1 and 4.2), and they are joined at seams that leave grooves (sulci) in the underlying bone. Scute terminology is not standardized but most authors follow the names proposed by Zangerl (1969). Anteriorly along the centerline, some tortoise species have a small cervical or nuchal scute, but this is frequently absent in the Texas tortoise. Posteriorly along the centerline are 5 uniquely shaped vertebral scutes. Along the side and touching the vertebrals are 4 pleurals, which laterally touch the 23 marginals. There are 11 marginals along each side and a single large posterior one.

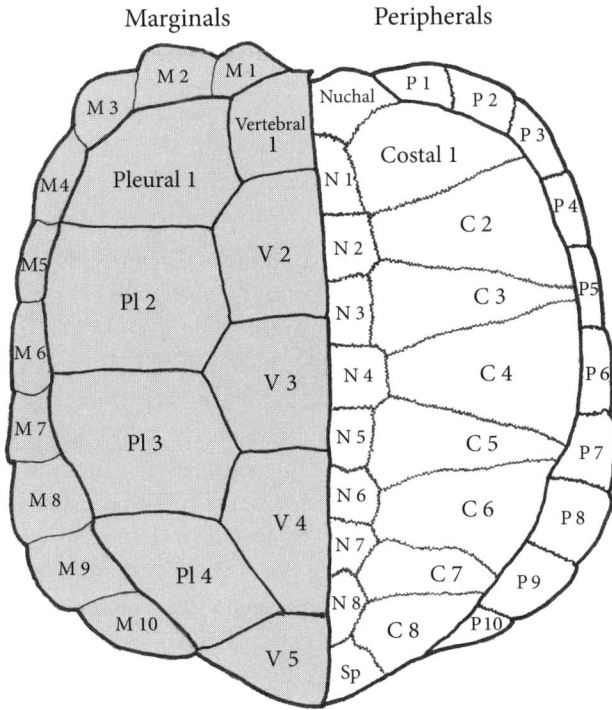

Marginals   Peripherals

Figure 4.1. Carapace scutes (left) and bones (right) of a
Texas tortoise. Because the tortoises are dome shaped, it
is difficult to capture all anterior and posterior elements
in a single illustration. Scutes: M = marginal, Pl = pleural,
V = vertebral. Bones: C = costal, N = neural, P = peripheral,
Sp = suprapygal.

Figure 4.2. Posterior carapace scutes
(left) and bones (right) of a Texas
tortoise. Scutes: M = marginal,
Pl = pleural, V = vertebral. Bones:
C = costal, N = neural, P = peripheral,
Sp1 and Sp2 = suprapygals 1 and 2.

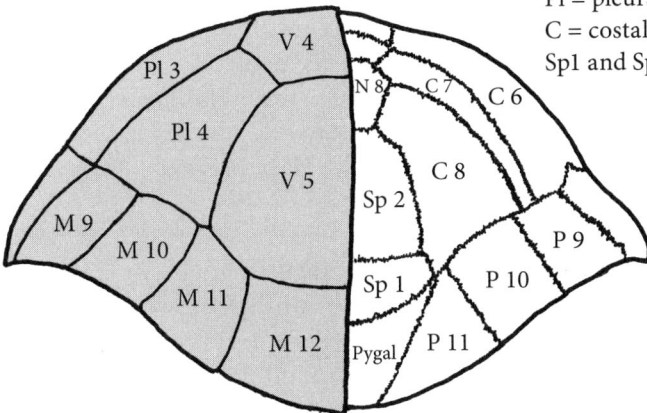

Plastron.   The plastron (fig. 4.3) is composed of four paired bones—the epiplastrons, hyoplastrons, hypoplastrons, and xiphiplastrons, which form a ring around the ventral periphery of the tortoise—and a single element, the entoplastron, which binds the anterior bones together. The entoplastron gets little attention in the literature, but its internal posterior elongated spine makes it structurally interesting. The plastral elements are thought to be derived from the clavicles, interclavicles, abdominal ribs, and possibly other bones. The epiplastrons are homologous to the clavicles of other amniotes (Zangerl 1939). Although the posterior plastron of members of the genus *Gopherus* was thought to be rigidly fixed until 1991, a hinge occurs between the hypoplastron and the xiphiplastron that functions in females to lower the posterior plastron and facilitate egg passage, and in males it may allow the carapace-plastron opening to accommodate penis extrusion (Rose and Judd 1991; Barton 2006) (see fig. 1.6). Richmond (1964) pointed out that the plastron has three basic functions: (1) to protect the tortoise, (2) "to serve as a tension member that holds the lateral edges of the arched carapace in position," and (3) "to provide a ventral anchorage for the pectoral and pelvic limb girdles."

There are five pairs of plastral scutes (fig. 4.3), separated along the centerline by a seam. The gulars cover the bony extension of the epiplastrons and are followed posteriorly by the humeral, pectoral, large abdominal, femoral, and anal scutes.

There is sexual dimorphism in the shells of adult Texas tortoises. Notably, in males the gular extensions are longer and may be forked, the plastral concavity is deeper, and the bony elements are thicker, especially at the junction of the posterior plastron and bridge (see fig. 1.7).

There has been considerable discussion regarding the evolutionary development of turtle shells. Two primary theories have risen to the top of the heap: the Plastron First theory posits that the plastron formed first and then bony extensions were added to the ribs and completed the carapace. The Osteoderm theory posits that the shell developed from bony plates called osteoderms that were embedded in the skin. The recent discovery of a 220-million-year-old fossil turtle (*Odontochelys semitestacea*) in China (Li et al. 2008) may have given the edge to the Plastron First theory, because this fossil has a well-developed plastron and no carapace elements or osteoderms. The mills of paleontologists grind slowly but they grind exceedingly fine. For those interested in the conundrum of how the turtle shell evolved, we suggest Nagashima et al. (2009) and Rieppel (2009).

Skull.   The skull is composed of three sections: (1) the *cranium*, or brain case, which includes the upper jaw and the apparatus by which the lower jaw attaches to the skull; (2) the lower jaw, or mandible; and (3) the hyoid complex.

There are 21 named bones that make up the skull, 17 of which are paired and 4 of which are single elements; thus, there are 38 bones total. The single ele-

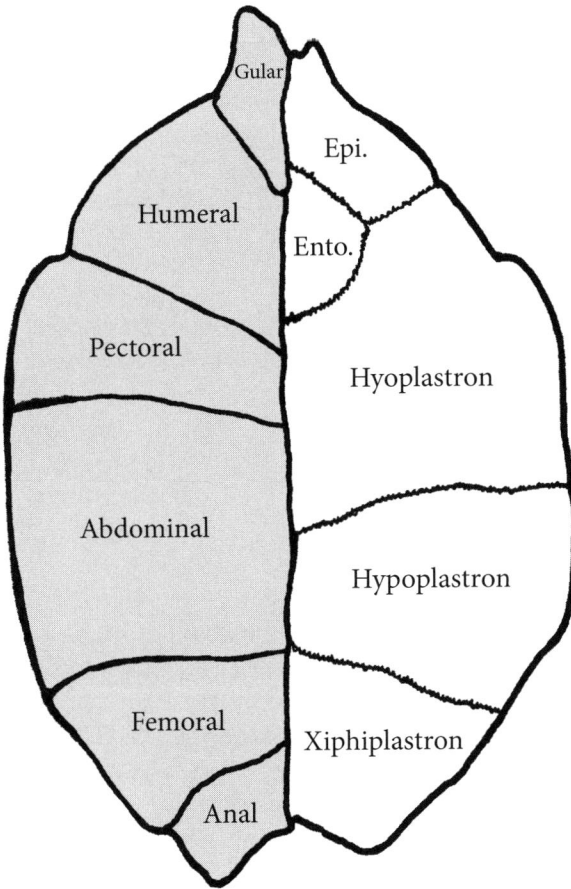

Figure 4.3. Plastral scutes (left) and bones (right) of a
Texas tortoise. Ento. = entoplastron, Epi. = epiplastron.

ments are in the central roof of the buccal (mouth) cavity and are clearly visible.
The two tiny columellae (associated with hearing) are normally not listed with
the bones but they are easily observed in the auditory cavity. These tiny but
conspicuous elements are probably a homologue of the stapes in the middle ear
of mammals.

On inspection, the brain case is noted to be small and the eye and nasal
cavities large (fig. 4.4). There are two large openings posterior to each eye, and a
large scalloped area between the jaw region and the otic (hearing) area. Auffenberg (1976) noted that the skulls of males are relatively longer than those of females, and that skulls of all members of the genus *Gopherus* are a mix of primitive and specialized characteristics. For example, the median premaxillary
ridge is shared with a fossil ancestor (*Stylemys*), the exposed prootic is primitive,
and the os transiliens (see fig. 1.1) is known only in the five extant species (Ray

Figure 4.4. Auffenberg's drawings of the bony elements in the skull of a Texas tortoise. The detail in these drawings is exceptional. After Auffenberg (1976); reproduced by permission of the Florida Museum of Natural History.

1945; Legler 1962; Patterson 1971a; Bramble 1974), but it undoubtedly also occurs in the Morafka desert tortoise. The prefrontals are short, the postorbitals are reduced, the quadrate encloses the columella, and the maxillary has three ridges—all specialized characteristics. The three ridges on the maxillary provide the skeletal core of the three cutting edges observed in the cornified covering of the upper jaw. The inner palatine ridge is the lowest and the outer is the highest; the medial ridge is intermediate in height and incomplete anteriorly.

Mandible.   The mandible attaches posterior to the skull at the mandibular condyle to form a hinge joint. Each side, called a ramus, is fused anteriorly at the symphysis and is composed of six bones. The major portion of each ramus is composed of a single bone, the dentary, which has a deep, longitudinal groove that accepts the middle ridge of the maxillary and is covered in horn; the horn provides the cutting surface to shear tough grasses. The outer supra-angular and inner prearticular form a cup posteriorly that accepts the dollop of bone called the articular, which articulates with the quadrate of the mandibular condyle. A thin sliver of bone, the angular, forms the posterior ventral surface of the mandible and on occasion participates in the cup that houses the articular. The coronoid is a small interconnecting bone that ties all but the angular together to form a peak that serves as a site for powerful muscle attachments. There is a well-defined Meckelian groove on the inner surface of the ramus that houses Meckelian cartilage, which emanates from the Meckelian foramen. A splenial bone is absent.

As strange as it appears, the hyoid apparatus is a portion of the skull. It forms the half basket of bones that supports the throat musculature; it allows a turtle to pump air and water and is important for respiration and olfaction. The primary element (basihyoid) in the Texas tortoise is shaped like a Y (plate 19) with a broad base. The dorsal anterior-posterior groove in the basihyoid houses the trachea. There are two paired bones that sweep upward and backward. The anterior elements are called anterior cornua and the posterior elements are called posterior cornua (the noun "cornu" refers to a hornlike projection). The anterior cornua are formed embryologically from the first branchial arches, and the posterior cornua from the second branchial arches. Taken out of their anatomical context, the hyoid apparatuses of various organisms, including the Texas tortoise, appear as mesmerizing and beautiful shapes. Within the human hyoid, which facilitates vocalization, the primary elements of the basic design are evident. In all animals with hyoids, the hyoid bone(s) are the only bony element(s) not attached to other bones.

Cervical vertebrae.   There are eight cervical vertebrae, although the first (the *atlas*) (fig. 4.5) is highly modified and looks nothing like the others. It is thin, fragile, and complex and occurs in five parts, loosely joined together. Three elements form a ball and socket joint that receives the occipital condyle of the skull. This unit is open at the top, that is, it does not form a complete ring. Another element is a small U-shaped bone attached to the ventral segments of the atlas and the first cervical vertebra. These elements provide little structural support but facilitate movements of the head. A second cervical vertebra (*axis*) may be attached firmly to the third. The eighth attaches securely to the first thoracic vertebra of the carapace via three surfaces. Lateral and vertical motion is somewhat limited by this arrangement, but there is not much need for these kinds of movements. The length of the neck allows it to move when extended, and its curvature allows the head to be withdrawn completely; however, the confines of the carapace and plastron and the shape of the cervical vertebrae place functional limits on the degree of movement.

Thoracic vertebrae.   There are eight, sometimes nine, thoracic vertebrae (plate 20) in the Texas tortoise. The second to eighth vertebrae are compressed and anchored to the neurals by a thin sheet of bone, and the eighth vertebra is the first typical vertebra with lateral extensions (transverse processes). At the junctions of the thoracic vertebrae is a thin sliver of bone that attaches laterally to the center of a neural, but at the eighth neural the structure becomes more complex. This thin sliver is thought to be a rib head extension that has been incorporated into the neurals. There are 10 to 12 sacral vertebrae, which are distinguished by stout transverse processes that slope backward and away from the spine at 130°. They are generally attached to the bony knob on the last costal, which supports the pelvic bones. Undoubtedly, there is some rib head involvement with these lateral extensions. The caudal (tail) vertebrae vary in number with the age and sex of the tortoise (males have a longer tail), but 15 is a good round number.

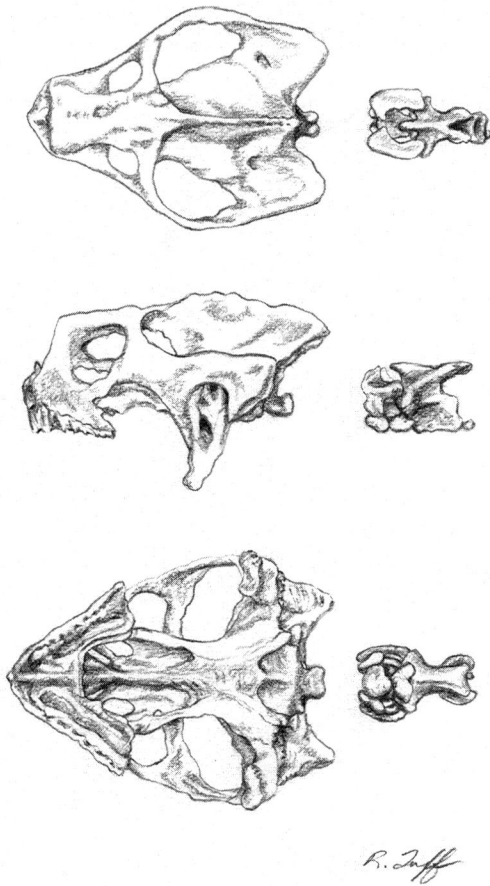

Figure 4.5. Dorsal, lateral, and ventral images of the skull and the atlas and axis complex that anchors the cervical vertebrae to the skull. Note the small singular knob on the skull that is encapsulated by the highly modified atlas.

They are typical in shape, and transverse processes may be well developed or minuscule. The first three may be attached to the carapace by connective tissue; the rest are free, allowing the tail to fold laterally.

The speed of neuronal impulses is directly related to the diameter of the nerve tract; thicker fibers carry impulses at a faster rate than those of lesser diameter. The small diameter of the spinal cord in the thoracic vertebrae implies that speed is not an essential component of Texas tortoise behavior. The canal, however, is enlarged starting at around the seventh neural. The peripheral spinal nerves typically pass from the spinal cord to the appropriate organs via well-developed spaces between the vertebrae. These openings (intervertebral foramina) are small and sporadic in position in the thoracic vertebrae of Texas tortoises but are clearly present in sacral, caudal, and cervical vertebrae. Clarification of this aspect of Texas tortoise anatomy through comparative studies with other turtles would be of value. While the tortoises' spinal cord anatomy suggests that a lot of their behavior is neuronally based in spinal reflexes, it also signals us that tortoises are built for the long haul; and, until the advent of humans and their vehicles, the tortoises' paradigm of "slow is better" was adequate. They live now in a fast-paced world, a world with ever-increasing human access to their habitat, a world where tires do not distinguish between a tortoise and a bump in the road. Thus they find themselves in the midst of technological marvels even as they are being slaughtered on the roads.

Appendicular Skeleton

Pectoral region.   Of the two girdles that serve as leg attachments, the pectoral is the least anchored: it has no firm attachments to the plastron or carapace. It has three elements. The scapula is a long, dorsally projecting rod that attaches in the region of the first thoracic vertebra and its associated rib. This ligamentous attachment often contains a sesamoid (extra) bone. Ventrally, the scapula has what appears to be a right-angle extension that projects anteriorly toward the midline; however, this extension, known as the acromial process, is a solidly fused bone, the precorocoid. The corocoid is a flat, bladelike bone that attaches to the scapula-acromial process complex and extends posteriorly toward the midline. Where the three elements converge, they form a socket, the glenoid fossa, which accepts the head of the humerus.

The humerus connects at its distal end (the end most distant from the center of the body) to two elongated bones, the slender radius and the larger ulna. Both are cupped securely around the distal end of the humerus and attach to carpals. The historical (phylogenetic) trend in tortoises has been toward fusion of carpals, and the trend during a lifetime (ontogenetic) is also for certain elements to fuse (fig. 4.6). In addition, we have difficulty assigning names to the carpal

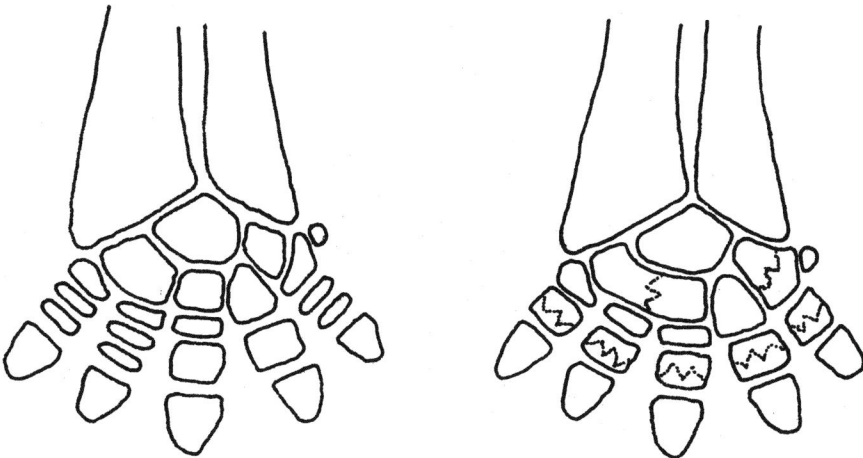

Figure 4.6. The carpal elements of Texas tortoises; a young animal on the left, an older one on the right. Note the fusion of the elements with increasing age of the tortoises. After Auffenberg (1976); reproduced by permission of the Florida Museum of Natural History.

elements in the limited number of samples we have, and so we refer readers who desire further information to the detailed treatments of Auffenberg (1976) and Crumley (1994). There are five named carpals (intermedian, ulnare, medial centrale, lateral centrale, and pisiform), and five smaller, more uniformly shaped ones, each of which is attached to one metacarpal per digit. The two outer digits have one phalanx (terminal bone) each and the three inner ones have two phalanges. The pisiform is a tiny, frequently overlooked bone that is attached laterally to the fifth carpal.

Pelvic region. The pelvic girdle is composed of two halves, each with three bones. The ilium is the vertical pelvic strut that connects to the last costal of the carapace. This is an active joint, as can be seen when the carapace bone is frequently worn away at the attachment site (see plate 18). Each anterior-projecting, triangular-shaped pubis connects medially to the other pubis. A thick triangular cartilage called the epipubis extends forward from the pubes, and a large pectineal process extends outward and forward. The posterior ischium has a backward projection and a strut, which along with a strut from the pubis and ilium forms the acetabulum, or socket, for the head of the femur. A large hole, the obturator foramen, is formed by the pubis and ischium. This foramen, the pectineal process, and the projection from the ischium are important sites for the attachments of muscles governing leg movement. The two ilia and the ventral floor of the pelvic girdle formed by the ischium and pubis form the pelvic canal, and it is important to remember that intestinal, reproductive, and urinary products must exit through this portal, which has considerable muscle mass attached internally. In some egg-bound tortoises, this portal is simply not large enough to accommodate the passing of large hard-shelled eggs (see fig. 1.5 and the discussion of dystocia in chapter 3).

Again, the structure of the tortoise's hind limb is standard for a vertebrate: a single femur is socketed in the acetabulum of the pelvic girdle and attaches distally to two bones, the larger tibia and smaller fibula. The foot, or pes, has to be considered in a vertical context because the hind foot is like that of an elephant; thus, it is difficult to visualize the foot bones as they occur intact and in space. The postradial elements are called tarsals, three of which (the tibiale, intermedian, and centrale) fuse to form one bone that accommodates the distal ends of the tibia and fibula. The fibulare is usually a distinct element. Auffenberg (1976) indicated that there are five tarsals but that the fourth and fifth are occasionally fused. Supposedly, there are five metatarsals (bear in mind that the fifth toe of the tortoise is not externally visible and is represented internally by a thin, square wafer of bone). There are generally two phalanges on the first to the fourth digits but only one associated with the fifth digit. Again, readers who desire more detailed knowledge about the foot elements may consult Auffenberg (1976). Because the foot bones of all four species of *Gopherus* are so similar, they have received little attention. The operative words when evaluating the carpal and tarsal elements of tortoises are like those used in plant identification keys: "usually," "sometimes," and "frequently."

Viscera

Heart

Texas tortoises, like all tortoises, have three-chambered hearts. The heart is composed of two auricles and a ventricle, along with the pipes and valves necessary to move blood along. Atrial cavities inside each auricle are separated from each other by an interatrial septum. The atrium of the right auricle is larger than the atrium of the left auricle. The auricles are relatively thin walled compared to the ventricle, and the wall of the left auricle is thinner than that of the right one. The sinus venosus is a relatively large chamber attached to the dorsal surface of the right auricle. Its wall is muscular, but even thinner than the auricle walls.

The ventricle is the most muscular chamber of the heart, and its internal wall has numerous spongy trabeculae. Its cavity volume is small and subdivided by incomplete trabeculae into three semichambers, or cava. From left to right, these are the cavum arteriosum, cavum venosum, and cavum pulmonale. Because the three ventricular cava communicate and contraction occurs in all three cava simultaneously, oxygenated and deoxygenated blood mixes in the ventricle and exits into all three arterial trunks at the same time. While we might consider this pumping system inefficient, it suits tortoises just fine. Even crocodiles and alligators, the only reptile groups to have four-chambered hearts, still send mixed oxygenated and deoxygenated blood to the main body. Birds and mammals are the only vertebrates that have cleared the anatomical hurdle of partitioning the chambers in such a way that oxygenated and deoxygenated blood remains separated in two separate ventricles. These anatomical changes made possible the spectacular physiological advances associated with higher blood pressure, which allows the more efficient transport of oxygen and carbon dioxide. Further, they allowed the development and control of a high-pressure renal system that coevolved with the refinement of a stable body temperature to allow better homeostatic maintenance of the internal milieu.

Heart rates of Texas tortoises are exceptionally low and are impacted by temperature. At room temperature their hearts beat only about six times per minute. The rate is higher at any given temperature when the animal is warming than when it is cooling (fig. 4.7). When tortoises were acclimated to temperatures as high as 35°C, the mean heart rate was about 22 beats per minute (Rose and Judd 1982) (fig. 4.8).

Alimentary (Digestive) Canal

The anterior opening into the alimentary canal is the mouth, or buccal, cavity. Extant tortoises have no lips or teeth, but they are equipped with well-developed tongues that are not protrusible. A small Jacobson's organ (vomeronasal organ)

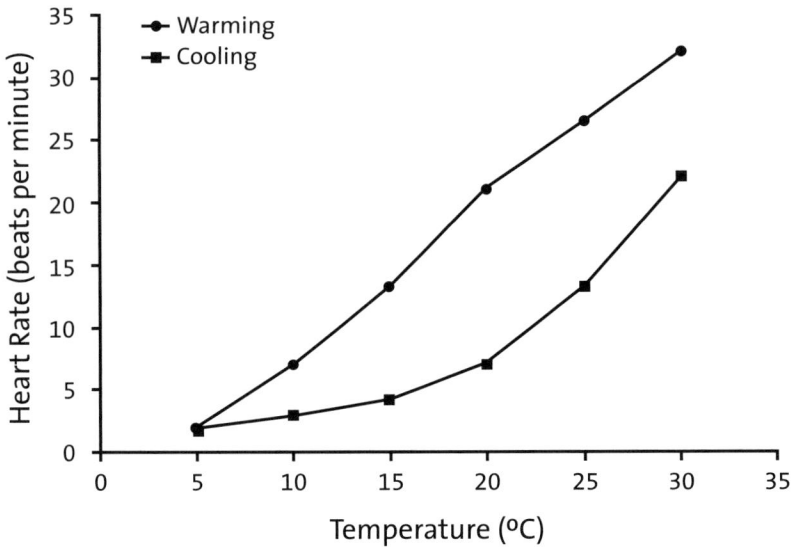

Figure 4.7. Heart rates of Texas tortoises while warming and cooling. The lower heart rate when cooling delays heat loss, and the higher rate facilitates heat gain while warming.

has an opening just inside the mouth, and immediately posterior to it are the paired internal nares. Jacobson's organ is a chemoreceptor and is probably involved with the detection of reproductively important pheromones. There is no hard or soft palate; thus it would be difficult for a tortoise to eat and breathe at the same time.

The pharynx is a short passageway posterior to the buccal cavity. An epiglottis valve is present ventrally in the pharynx and marks the entrance to the trachea of the respiratory system. The alimentary tract continues dorsally as the esophagus, a distensible, muscular-walled tube between the pharynx and the stomach. The eustachian tubes, one on each side, which equalize pressure in the auditory cavity, open into the buccal cavity through the roof of the pharynx. The position of the entry to the respiratory system (epiglottis) below the opening of the esophagus does not bode well for proponents of intelligent design.

The stomach is a thick-walled tube that is markedly larger in diameter than other portions of the alimentary tract. It is slightly hooked at the posterior end, which causes its overall shape to resemble the letter *J*, and it narrows markedly at its distal end to a thick muscular sphincter (ring of muscle) called the pyloric valve. The stomach may be noticeably larger than that of meat-eating turtles. However, there are as yet no comparative studies of the sizes of the alimentary canal components of tortoises and those of other turtles with different feeding paradigms. The pyloric valve controls the movement of the food bolus into the small intestine, which is a long, narrow, coiled tube with little external or inter-

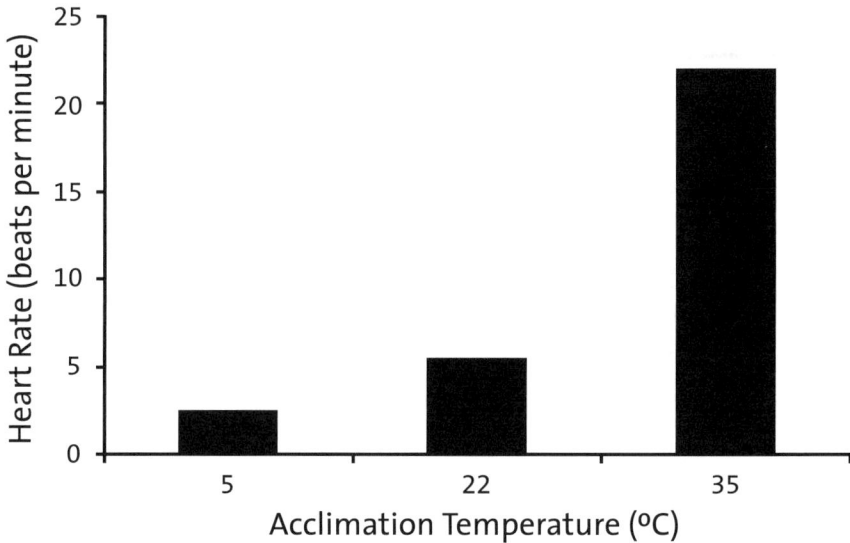

Figure 4.8. Heart rates of Texas tortoises that were acclimated at three temperatures. Acclimation lowers the heart rate significantly below that of unacclimated individuals.

nal regional differentiation. The gall bladder and pancreatic ducts empty into the anterior portion of the small intestine, which is referred to as the duodenum. The transition from the small to large intestine is abrupt and easily recognized, as the diameter of the large intestine is markedly greater than that of the small intestine. The large intestine is the least muscular and most thin-walled portion of the alimentary tract, and unlike the small intestine, it is straight. An ileocolic sphincter is present at the end of the small intestine where it joins the large intestine. This sphincter regulates the passage of food from the small to the large intestine. A small outpocketing called the caecum lies adjacent to the juncture of the small and large intestines. The cloaca is a common passageway for the digestive, reproductive, and urinary systems. The anus is a puckered vent through which cloacal products exit the body.

A number of organs and glands are associated with the alimentary canal and digestion. Numerous unicellular and multicellular glands are present in the interior lining of the alimentary canal. Most of these secrete mucus, and some secrete digestive enzymes or acid. The liver is the largest of the accessory digestive organs and serves a number of functions including glycogen storage; blood cell formation; detoxification; the production, assembly, and degradation of various substances; and bile production. Bile drains via numerous small ducts from the liver into the gall bladder, which is embedded in the surface of the liver's right ventral lobe. From the gall bladder, bile drains into the small intestine, where it emulsifies fats, thereby aiding digestion. One might question the function of bile in an herbivore, but plants also contain fats. The gall bladder is also a waste receptacle that accepts and stores degradation products from old, injured, or malformed red blood cells that are rendered by the liver. The digestive function was evolutionarily conscripted as an aid to digestion when the transition from herbivory to predation occurred.

The pancreas is much smaller than the liver and more diffuse. It is embedded in the mesentery supporting the intestines and lies adjacent to the duodenum. It secretes digestive enzymes and produces the hormone insulin. We know very little so far about the general digestion of Texas tortoises.

Respiratory Tract

Air can enter the respiratory tract through the external nares or through the mouth. Air entering from the external nares has a short passage, as the internal nares are located in the anterior of the buccal cavity. From the buccal cavity, air passes to the pharynx and from there to the epiglottis, which is the opening into the larynx. The larynx is short and opens into the trachea, which is easily recognized by its cartilaginous rings. The trachea divides into a right and left bronchus, and each of these subdivides into bronchioles that terminate in tiny alveoli, sacs with walls only one cell thick. These sacs are nestled in blood vessel complexes and are the sites where respiratory gases, notably oxygen and carbon dioxide, are exchanged. The lungs consist of bronchioles, alveoli, blood vessels, and supporting tissues. They have a complex pattern of interior folds and subdivisions that increase surface area and give the lungs a spongy consistency.

Turtles lack a diaphragm that separates the pulmonary and abdominal cavities, and they also lack movable ribs, both of which play key roles in mammalian respiration. Envision a closed box with one port. A tube (trachea) is inserted tightly into the port and an uninflated balloon is attached on the stub inside the box. To fill the balloon, you can (1) apply external pressure that is greater than the pressure inside the box (blow into the balloon), or (2) increase the space inside the box and thus decrease the internal pressure. However, the trick in respiration is not to empty the balloon, but to keep it partially filled at a specific internal pressure to maximize oxygen and carbon dioxide transfer. The tortoise's lungs are tightly attached to the underside of the carapace, where they are separated from the other organs by a sheet of tissue called the diaphragma. In addition, a tough membrane that connects to the diaphragma surrounds the internal organs. Two antagonistic muscles act on this membrane to help increase or decrease the pressure within the body cavity.

Transverse abdominal muscles attach to the inner carapace, and when they contract they pull against the diaphragma, which pushes on the lungs, and air is expelled (expiration). Oblique abdominal muscles attach to the skin of the hind legs, and when contracted they pull the membrane connected to the diaphragma back like the plunger in a hypodermic syringe, decreasing the pressure and allowing air to enter the lungs (inspiration). The shoulder girdle is also

Plate 1. Juvenile Texas tortoise showing yellow coloration. Note the all-yellow head. Image used with permission of Robert and Linda Mitchell, Robert Mitchell Photography.

Plate 2. *Manouria emys* (Asian brown tortoise), closest living relative (sister group) to members of the genus *Gopherus*. It is found from Assam in extreme northern India to Malaysia, Sumatra, and Borneo.

Plate 3. Subdentary glands of a male Texas tortoise. Note that one gland is oozing fluid. Also, note the eye color and the membrane (tympanum) over the hearing apparatus.

Plate 4. The forelimbs of Texas tortoises protect the head from predators.

Plate 5. The hind feet of Texas (and other) tortoises protect the inguinal regional from predators. Note the short, small tail of the female.

Plate 6. Entrance to a pallet constructed by a Texas tortoise. Note the classic half-sphere appearance of the opening.

Plate 7. Thorn-shrub–grassland community.

Plate 8. Loma and cactus-grass community.

Plate 9. Arroyo community.

Plate 10. Mesquite in a savanna grassland community.

Plate 11. Fire damage on the Texas Parks and Wildlife Chaparral Wildlife Management Area (WMA) in South Texas. The fire in 2008 devastated 95% of the 15,200-acre area, one of only a few places where Texas tortoises have been studied extensively. Used with permission of the Chaparral WMA, Texas Parks and Wildlife Department.

Plate 12. Cactus (*Opuntia lindheimeri*, or *O. engelmannii*) with ripe red tunas, a frequent food of Texas tortoises.

Plate 13. Forelimb injury of a Texas tortoise named T-bone, induced by a raccoon or fox. The injury as seen here is in a state of healing after intense medical intervention. The radius and ulna ultimately fell out, as did the distal bulb at the end of the humerus, and the injury quickly closed. Although the leg is misshapen, locomotion is not seriously impacted and neither is T-bone's indomitable spirit (see plate 30).

Plate 14. Result of a bone infection in a Texas tortoise named T-bone. These infections are not easily cured with medical intervention. In this case, the progression ceased when it reached the soft tissue, where increased blood flow and immunological stopgaps were presumably in place.

Plate 15. Texas tortoise (Mashed Morris) exhibiting severe carapace damage, probably from a vehicle encounter. How the spinal cord and internal organs survive such injuries is not understood.

Plate 16. Remnants of fire-induced injuries on a Texas tortoise.

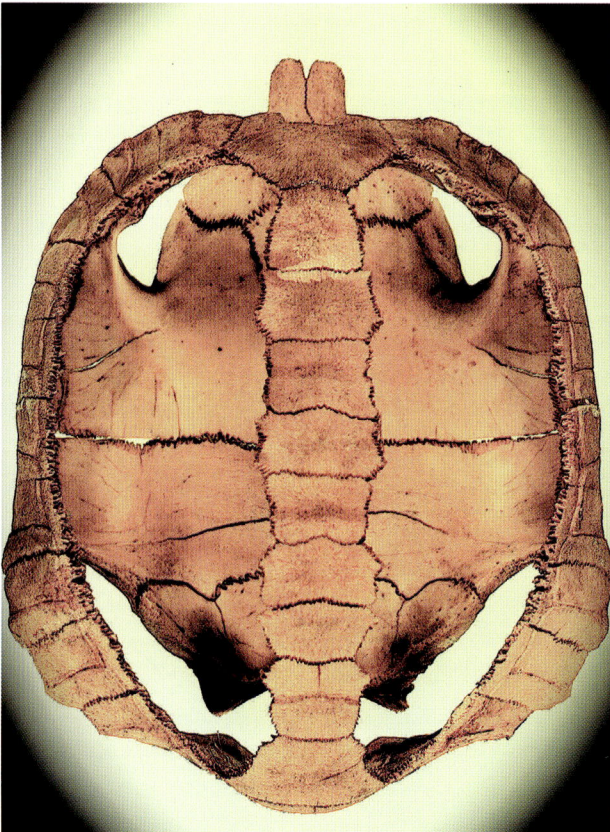

Plate 17. Carapace bones of a Texas tortoise with the costals removed, demonstrating how much of the carapace is composed of incorporated rib elements.

thought to enhance respiration (which may account for its loose attachments to the carapace and plastron) as the legs are extended and withdrawn. Any sudden withdrawal of the forelimbs of the Texas tortoise is accompanied by a short hiss as air is expelled. Often, when Texas tortoises breathe deeply, or when they are in respiratory distress, they extend the neck fully, open the mouth, and visibly expand the throat as air enters. Thus, they can also use the tried and true method of respiration inherited from their amphibian ancestors: draw a volume of air into the mouth and throat regions, close off the external nares, seal the mouth, and "swallow" the air into the lungs.

The buccopharyngeal area of Texas tortoises can be seen rising and lowering up to 20 times per minute. These movements are probably olfactory in function. Depending on the temperature and the tortoises' activity level, the actual breathing cycle of inspiration and expiration occurs every two to four minutes, but if tortoises are exposed to noxious fumes, such as chloroform, they may hold their breath for an hour or more.

Urinary Organs and Tract

The two small kidneys have lobular surfaces and lie side by side, one on either side of the vertebral column on the dorsal body wall in the coelomic cavity. A ureter drains urine from each kidney to an elastic-walled urinary bladder. The thin-walled, bilobed urinary bladder joins the cloaca, from which it forms during development, through a short median duct called the urethra that transports urine with its uric acid and urate salt components. Lacking loops of Henle in the kidneys, reptiles cannot retain water by producing urine more concentrated than the blood plasma. Uric acid works well as a urinary product for organisms living in a hot, dry environment because it occurs in crystal form and allows the elimination of two ammonia molecules for the price of one. In addition, because uric acid is an osmotically inactive molecule, water can be passively reabsorbed from the urinary bladder, colon, and cloaca. Chelonians differ from other reptiles in that the ducts draining the kidneys enter the neck of the bladder rather than its wall. True (1881), while preparing a desert tortoise specimen, commented on finding "on each side of the body, between the flesh and carapax, a large membranous sac filled with clear water; I judged about a pint run out." True also observed, "He carries his supply of water in two tanks. The thirsty traveler, falling in with one of these tortoises and aware of the fact, need have no fear of dying of immediate want of water." Auffenberg (pers. comm.) commented that the bladder in desert tortoises was nearly twice the size of that in comparably sized gopher or Texas tortoises and was positioned much as in True's description. He surmised that the size and position of the bladder (in contact with the carapace) might allow it to serve as a thermal reservoir. Certainly, the bladder of the Texas tortoise does not reach the size implied by True's description, even when scaled to its body size. Woodbury and Hardy (1948) made no mention of the position of the bladder, other than what would be expected, but they did draw the bladder as a bilobed structure of significant size. Regardless, we think that the glob of uric acid crystals and the slimy constitution of the bladder fluid (urine) of the Texas tortoise might give pause to even the most drought-stricken traveler.

Reproductive Organs and Passageways

The reproductive systems of male and female tortoises are typical of those of other reptiles. Reproductive products, urine, and feces all exit through a common port, the cloaca. While musing on this staple of reptilian anatomy, one can but wonder what fickle designer contemplated such a loathsome contrivance that so firmly joins the pleasure apparatus with the septic system. It is not known how tortoises or other reptiles, especially females, protect themselves from the pathogenic microorganisms likely to be introduced to their systems from the mixing of waste and reproductive products in the cloaca.

Male system.   Testes are paired and lie adjacent to the anterior end of the kidney. Each testis is a mass of seminiferous tubules, interstitial cells, and blood vessels encased in a connective tissue sheath. The walls of the seminiferous tubules are lined with germinal cells that produce spermatozoa. Sperm produced in these tubules pass through an efferent duct into the epididymis on the medial surface of the testis. The epididymis delivers sperm via the ductus deferens to the base of the penis. The testes undergo marked seasonal fluctuation in size, being largest when spermatogenic activity is at its peak. Similarly, the epididymis and ductus deferens vary in size seasonally.

The penis arises from the floor of the cloaca. Jones (1915), who suggested that studying the various degrees of penis development in turtles might lead to an understanding of the evolution of the mammalian penis, was the first to make an organized attempt at such a study. Zug (1966) evaluated penile structure in turtles relative to phylogeny. He did not describe the penis of the Texas tortoise (although he had three specimens) but covered the structure under the umbrella of *Gopherus*. When engorged, the penis of the Texas tortoise appears unnecessarily complex (fig. 4.9), and rather impressive for the animal's size. On several occasions we have observed an extended penis being dragged across the ground behind a tortoise, only to see it slide back into the body cavity with a load of detritus. Its color is a brownish dull purple. The seminal groove delivers sperm to the terminal portion (glans); thus, there is no tube equivalent to the mammalian urethra that travels the length of the penis. The plica media, or corona, is a large, raised, horseshoe-shaped dominant feature. By now the reader should surmise that the penis does not serve to deliver urine from the body.

Female System.   Females have a pair of ovaries (fig. 4.10), each of which is an aggregation of epithelial cells, connective tissue, nerves, blood vessels, and germinal cell beds encased in an elastic tunic. An oviduct lies adjacent to, but is not

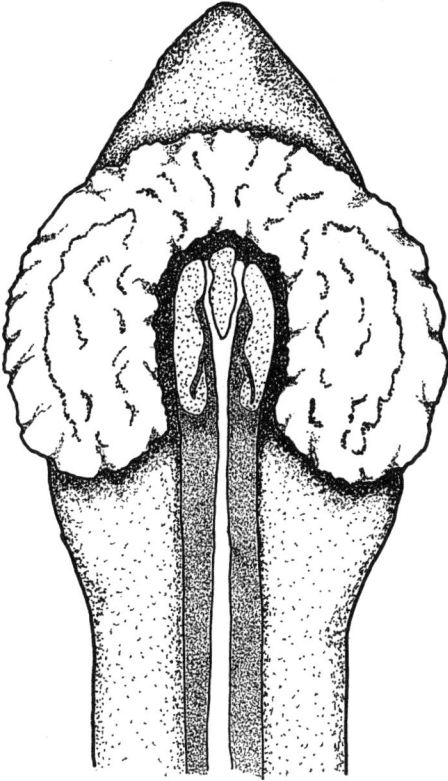

Figure 4.9. The penis of a young adult Texas tortoise. Note that the central groove is not a tube. The wall of the female cloaca serves as a cover under which the sperm can travel toward the tip. Redrawn after Zug (1966).

continuous with, the ovary. The opening into the oviduct (ostium) lies beside the anterior part of the ovary and enlarges during ovulation to entrap the ova as they are released. The body of the oviduct has an anterior albumin-secreting portion followed by a thicker, posterior, shell-secreting portion. The oviducts open separately into the urogenital sinus of the cloaca and vary seasonally in size in similar fashion to the male reproductive ducts. The cloaca of a female receives the penis of the male and serves as the roof of the seminal groove, allowing the sperm to reach the glans region. Because the seminal groove is not a tube, sperm delivery may be a rather leaky affair, and this may explain why copulation is sustained. There is a tiny protuberance, the clitoris, on the surface of the cloaca that is probably the homologue of the penis.

A large number of turtles are known to store sperm within the oviduct. In all groups studied (including the Texas tortoise), sperm storage tubules are located in the posterior albumen-secreting section of the oviduct (Gist and Jones 1989). The advantages of storing sperm are obvious on the surface; however, if courtship is necessary for ovulation, then some mechanism must be in place to trigger sperm activation. The location of the tubules housing the sperm at the posterior end of the albumen-secreting portion of the oviducts implies that

Colon

Clitoris

Cloaca

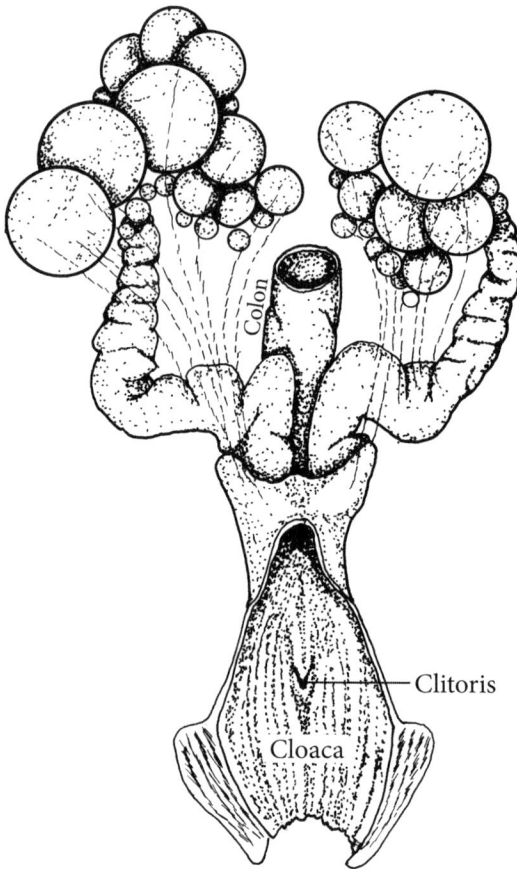

Figure 4.10. Reproductive system of a female tortoise. The ovaries and developing follicles are anchored by a fabric of membranes. Shelled eggs are not released into the cloaca until egg laying commences. Redrawn after Jones (1915).

sperm are released prior to the ova reaching the albumen-secreting portion. Fertilization would be more difficult, if not impossible, once the jellylike albumen surrounded the ovum. Vigorous courtship activity probably initiates ovulation and movement of sperm into, and up, the lumen of the oviduct, and fertilization occurs in the section preceding that of albumen production. Because of the sheer numbers and vigor of its sperm, a currently copulating male may have a higher probability of fertilizing the female's ova than a male that copulated with her previously, accounting for the long copulation times (up to 90 minutes). As mentioned under intrasexual behavior, pseudocopulation by females (see chapter 8) may be a mechanism to induce ovulation, thus providing stored sperm with access to unfertilized ova in the absence of males.

5

Size, Growth, and Sexual Dimorphism

A hen is only an egg's way of making another egg.
—Samuel Butler, cited in Matt Ridley, *The Red Queen:*
Sex and the Evolution of Human Nature

This chapter looks at the life history of the Texas tortoise, starting with a description of the eggs and their incubation and hatching. Parental care and growth are then discussed, as well as reproductive behavior and sexual dimorphism. Finally, the tortoise's variation in size, its longevity, and its survivorship are described.

Egg Size and Shape

Shelled (cleidoic) eggs are so commonplace that we scarcely give them thought, yet they were a supreme reproductive advancement. Other than the platypus and echidna, only reptiles and birds lay shelled eggs. While some lizards, snakes, and crocodilians defend the nest and its reproductive products, chelonians were thought to be exceptions in that once eggs were laid, the maternal parent abandoned them to their fate. But see the section on parental care later in this chapter.

Many amphibians require water to complete their development. The cleidoic egg of reptiles became its own little pond where embryos could develop somewhat and escape the vicissitudes of environmental perturbations. For some reptiles, but not for others, the entry of water into the eggs during development is necessary. A downside of shelled eggs that are laid outside the female's body is that if a predator finds one egg in a nest, it finds them all, and their high energy content makes them a choice target.

Turtle eggshells are composed of calcium salts and can be leathery or porcelain hard, round or elongated. To our knowledge, only Texas tortoises lay eggs that vary from nearly spherical to definitely elongated (plate 21), which reflects the number produced in a clutch (see below for explanation). Generalized components of an egg are: (1) a calcareous shell, (2) a collagen-based matrix membrane that separates the shell from its inner contents, (3) a protein-laden albumen, and (4) a yolk, the yellow core. All genetic and physical requirements to engineer a new individual are contained in this inauspicious and mundane housing. Unlike many turtle eggs, those of Texas tortoises, and most arid-adapted tortoises, do not require water for proper development.

The ovum (Latin for "egg") is configured as a sphere in the ovary. With the appropriate array and balance of hormones, the ovum is released from its follicle and travels to the opening (ostium) of the oviduct, where it is drawn inside by the sweeping action of microscopic cilia. Fertilization must occur at this stage, because albumen is quickly sequestered from cells lining the duct. Next, the fibrous matrix membrane is extruded around the albumen, and it is at this stage that egg shape is determined—small packets are spherical and larger ones are ellipsoidal. Calcium crystals are extruded into the matrix membrane as the product apparently spirals through the duct and comes to rest near the terminal end, where it awaits eventual expulsion.

Most Texas tortoise eggs are elliptical, with long (length) and short (diameter) axes. Egg length ranges from 37.0 to 46.9 mm (1.5 to 1.8 in), with an average of 41.5 mm (1.6 in), and egg diameter ranges from 30.0 to 36.2 mm (1.2 to 1.4 in), with an average of 34.1 mm (1.3 in). Egg mass ranges from 18.7 to 30.4 g (0.7 to 1.1 oz), with an average of 26.9 g (0.9 oz) (Judd and Rose 1989). Increases in both egg length and diameter result in an increase in egg mass. Egg volume ranges from 21 to 32 ml (0.7 to 1.1 fl oz), with an average of 28.3 ml (1.0 fl oz). In general, larger turtles within a given species lay larger eggs. This implies that optimum egg and clutch sizes are attained in older (larger) individuals. This is not the case for female Texas tortoises, where the pelvic aperture through which eggs must pass does not increase significantly in size ontogenetically. The pelvic girdle attains its maximum size in sexually immature females, as no significant growth is attained after sexual maturity (Long and Rose 1989). In addition, clutch size (number of eggs) does not increase with an increase in female size (fig. 5.1).

A regression of mean egg length on clutch size (fig. 5.2) shows a significant inverse relationship, but egg diameter is not related to clutch size. Thus, increases in clutch size result in eggs that are shorter in length, but not smaller in diameter (Judd and Rose 1989). Furthermore, egg diameters apparently cannot be increased past a mean of about 38 mm (1.5 in), which is determined by the size of the pelvic canal that houses muscle attachments and the cloacal walls.

In an example of the eyes being able to trump the brain, Grant (1960) and Paxson (1961) noted that the eggs of Texas tortoises, if hard-shelled, were too large to exit the carapace-plastron port. Grant based his statement on eggs observed after they were laid, and Paxson based his view on two eggs found inside a

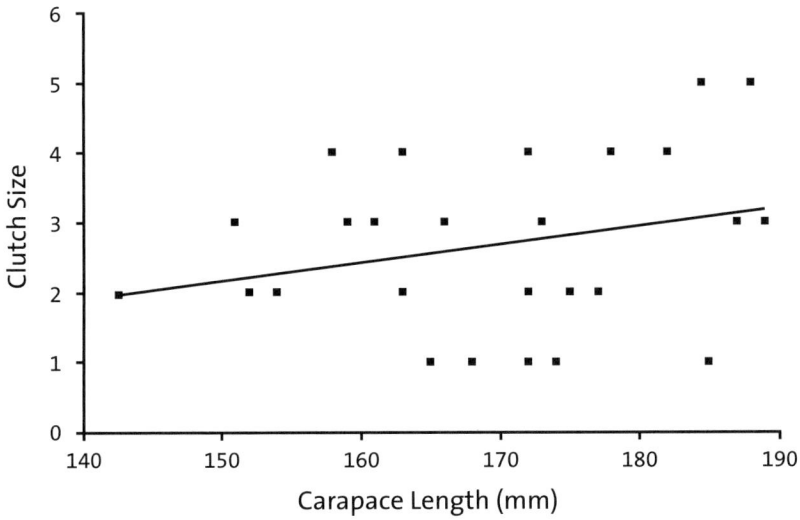

Figure 5.1. Clutch size versus size of female Texas tortoises. Clutch size does not increase significantly with the size of the female, indicating that clutch size is optimized.

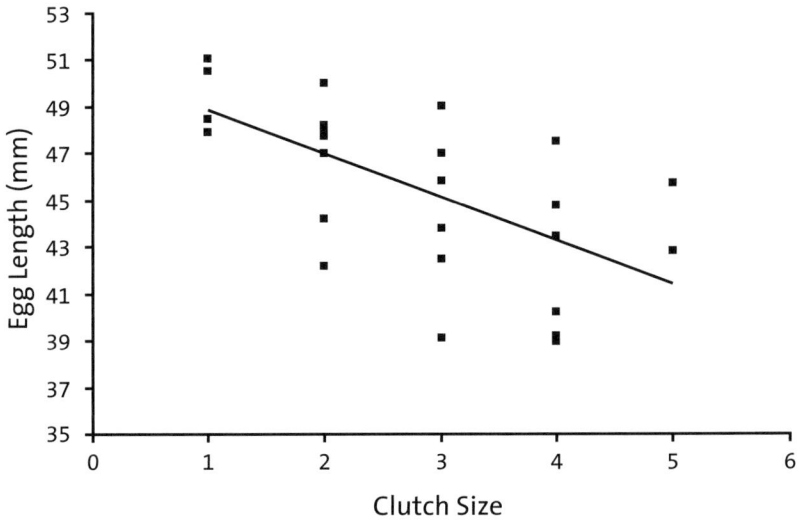

Figure 5.2. Egg length versus clutch size in Texas tortoises. The data confirm that average egg length decreases with an increase in clutch size, indicating that energy allotment to individual eggs is less with increasing egg length.

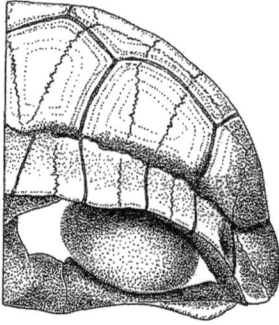

Figure 5.3. Demonstration of how the plastral hinge lowers and the posterior carapace rises to accommodate the passing of the large eggs of Texas tortoises (Rose and Judd 1991). Used with permission from the Society for the Study of Amphibians and Reptiles.

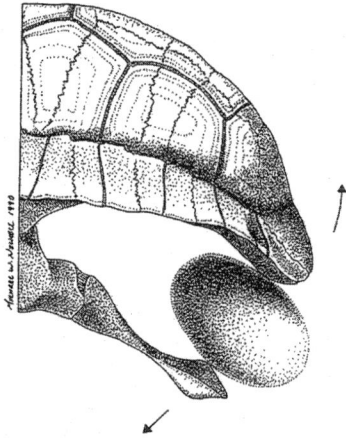

deteriorating tortoise shell. Paxson summarized "the self-evident but rather generally overlooked fact that the eggshells are flexible and flattened when laid, becoming hard and nearly spherical in a few minutes, and that the pygal-xiphiplastral distance could not possibly permit the passage of the hardened egg." Yet a plastral hinge is clearly evident in skeletal preparations without scutes and on x-rays. This is an excellent demonstration of reasonable and dedicated individuals failing to explore all possible explanations when addressing a conundrum, and from a faulty interpretation drawing two faulty conclusions about egg softness and shape.

We examined females killed on highways and found that eggs within the oviducts were hard. Also, we noted that eggs were hard shelled when laid and that the plastron and carapace were movable (see fig. 1.6) and could increase the space between the pygal and the xiphiplastron when the eggs were laid (Rose and Judd 1991). As an egg passes from the shell, the posterior carapace is forced in a posterior and slightly dorsal direction (fig. 5.3), and the xiphiplastron deflects ventrally.

The eggshell and membrane surround a packet of stored energy, water, chemical constituents, and genetic information that will be converted into an individual tortoise. It would be helpful to be able to hard-boil a tortoise egg to ascertain how energy is compartmentalized within it. That would enable the three primary components to be separated like those of a bird into (1) the shell and its inner matrix membrane, (2) the albumen fraction, and (3) the yolk. Estimates could then be made of the energy contained in each component. Unfortunately, no amount of boiling or microwaving hard-boils a tortoise egg. Because of this mysterious anomaly, it is best to treat the egg as having two fractions: the shell and inner membrane, and the yolk-albumen component.

When a Texas tortoise egg is broken open, the brightly colored yolk and the globular, glistening albumen are striking. In a freshly laid egg, if a cut is made along the shell's equator (plate 22), the albumen and yolk encapsulated in the upper half will remain in place when the cap is removed.

Throughout this book we use weight and mass interchangeably. Mass is a measure of the amount of matter in an object. Weight is the force that gravity exerts on that object. So, weight and mass are equal under the same gravitational loads. Nevertheless, scientific journals routinely demand weight data presented as mass. While we have weighed many things, we have never massed any.

The shell of an average-sized Texas tortoise egg weighs about 4 g, which is about 12% of the total weight of the egg. Yolk and albumen weigh about 28 g and make up about 87% of the egg's weight. Lipids make up about 2.2 g of the yolk and albumen and contribute about 6.9% of the total egg weight, while water makes up 69% of the yolk and albumen. It seems plausible that because Texas tortoise eggs are exposed to generally hot and dry conditions, a great deal of lipids and water would be stored in each egg. The amount of stored lipids and water might be slightly greater than what is found in eggs of freshwater aquatic turtles (Congdon and Gibbons 1985) that lay their eggs on land, but this is because of the large size of tortoise eggs; the percentages are similar.

Relative clutch mass (RCM) is the percentage of the postpartum weight of the female that is egg mass. Whereas this measure might have some relevance when attempting to evaluate reproductive output in various vertebrates, it is too variable in individual Texas tortoises to establish a meaningful statistic. For example, a female with a postpartum weight of 1210 g that laid four eggs (39.4, 29.1, 29.6, 26.2 g) had an RCM of 9.4%. If she had laid three eggs, the RCM would have been 7.0%, and for two eggs, 4.6%. Assuming average body size and a clutch size of three eggs, the average RCM for Texas tortoises is about 7%–8%. Smaller body sizes and the variable number of few large eggs render RCM data virtually useless as a comparative life history tool for this species.

Male tortoises also have loose articulations between the posterior carapacial and plastral elements, but they are not as loose and movable as in females (Rose and Judd 1991). This suggests that the loose posterior carapace and plastron elements are not an adaptation exclusively associated with egg laying and could perhaps facilitate respiration and storage of digestive matter and water. The looseness of these skeletal elements may also allow tortoises to increase pulmonary inspiration, thus increasing their size and making it more difficult for predators to extract them from pallets or burrow entrances. It is likely that the looseness of the posterior skeletal elements provides multiple adaptive advantages. To our knowledge, no one has investigated the action of the peptide hormone relaxin on the tortoise muscles that facilitate plastral hinge relaxation during egg passage.

Incubation

Incubation time for Texas tortoises ranges from 88 to 118 days (Judd and Mc-Queen 1980). Desert tortoises have a similar incubation time of 84 to 120 days (Jackson et al. 1976; Ewert 1979). Tortoises of the genus *Gopherus* may have similar incubation times regardless of species or adult size.

Duration of Egg Production

Auffenberg and Weaver (1969) reported that egg laying by Texas tortoises occurred from June 8 to August 2 and that freshly caught females sacrificed on September 16 had shelled eggs in their oviducts. They concluded that nesting dates in South Texas ranged from at least June 8 to September 17 and suggested that some eggs might be laid as late as November. Rose and Judd (1982) found a fresh nest on April 29 and reported that 11 of 13 females killed by vehicles in April 1972 contained shelled eggs. Thus, the combined information from these two studies suggests that egg laying begins in April and extends until at least mid-September.

Radiographs of females (see fig. 1.5) in 1986 and 1987 confirmed that calcification of eggs commenced in April but ceased in July in both years (Judd and Rose 1989). These results are similar to those of Turner et al. (1986), who found that in desert tortoises in California, shelled eggs were present from April 18 to July 4. Iverson (1980) reported that ovulation and oviposition of gopher tortoises from Florida occurred from late April to mid-July. Similarly, Landers et al. (1980) found that gopher tortoise nesting extended from May 18 to June 27. Thus, only Auffenberg and Weaver (1969) found oviposition occurring after July in a species of *Gopherus*.

The rainfall prior to and during the 1986 egg-laying season for Texas tortoises in South Texas was markedly different from that during the 1987 season (Judd and Rose 1989). Rainfall from August 1985 to July 1986 was 57.7 cm (22.7 in), and from August 1986 to July 1987 it was 84.3 cm (33.2 in). Differences in precipitation in the six months prior to the onset of egg laying were even more striking. Rainfall was three times greater in 1986–87 (50.8 cm [20.0 in]) than in 1985–86 (17.2 cm [6.8 in]). Given the similarity in the time of cessation of egg production in these two years, it seems unlikely that rainfall or food availability can explain the difference in cessation dates that we and Auffenberg and Weaver (1969) reported.

Hatching

Texas tortoises have an egg tooth, or caruncle, at the anterior tip of their horny beaks. Young are bent inside the egg, that is, there is a fold in the plastron and the head and tail are proximate to one another. Tortoises break free of the eggs in two ways: they use the caruncle to pip their way out of the egg in much the same way a bird does, or they straighten out from their flexed position near the end of incubation and cause a rupture in the eggshell along its length. Tortoises have often not absorbed all of the yolk when they hatch, so the yolk sac (plate 23) remains attached for several days until it is absorbed. Some individuals remain in the ruptured eggshell for several days.

In South Texas, eggs hatch from August 27 to November 5. Hatching success in 1978 was 60.0% and in 1979 it was 61.5% (Judd and McQueen 1980). Thus, there was little variation in hatching success between years. There are few data on hatching success for members of the genus *Gopherus*. Brode (1959) observed 40 gopher tortoise nests in Mississippi and reported that most of the five or six eggs in each nest hatched. Unfortunately, statistics that would facilitate comparisons were not given, but if "most" is interpreted as more than half, that is, four eggs from a clutch of six, or three eggs from a clutch of five, hatching success for gopher tortoises would be about 60% to 66%, similar to what we observed for Texas tortoises.

Hatchlings are soft and are mere popcorn to most predators. The sequence of shell hardening has not been addressed in Texas tortoises but was innovatively documented in the Agassiz desert tortoise (Nagy et al. 2011).

Auffenberg and Weaver (1969) reported that predation is heavy on Texas tortoise eggs. This may be because the last egg in a clutch may be laid on the surface, making it relatively easy for predators to locate the nest. However, nests are scattered throughout the habitat rather than being laid in communal laying areas, as they are for some turtles. Predators, then, do not have specific search areas for a high-energy meal and probably "stumble" onto nests during general prey searches.

Parental Care

Chelonians are the only group of vertebrates that do not typically exhibit parental care (recently reviewed by Agha et al. 2013). That is, once the eggs are laid, there is no further involvement with the parent(s), no nest protection, and no provisioning for the hatchlings other than what remains in the yolk sac. In their predator-driven world, chelonian young are on their own, but there are four caveats: (1) Iverson (1990) suggested that female yellow-bellied mud turtles (*Kinosternon flavescens*) may remain buried near the nest and possibly provide protection from predators and moisture to the eggs via urine (but see Burk et al. 1994); (2) female Asian brown tortoises (*Manouria emys*) are known to construct a debris nest on the surface and defend their eggs aggressively or passively for a few days by lying on top of the nest (McKeown et al. 1982); (3) female desert tortoises defend their nest tunnels from Gila monsters (Gienger

Table 5.1. Average size at hatching of eight Texas tortoises, size at one year of age, and instantaneous growth rates

Category	Weight (g)	Carapace length (mm)	Carapace width (mm)	Plastron length (mm)	Shell height (mm)
Initial size	24.3 (1.9)	42.2 (1.4)	41.5 (0.8)	39.0 (0.9)	24.6 (0.8)
Size at 1 yr	80.1 (6.3)	65.9 (1.3)	60.6 (1.1)	60.7 (1.3)	33.6 (0.9)
Mean increase	55.8	23.7	19.1	21.7	9.0
Mean % increase	229	56	46	56	37
Daily increase	0.15 (.014)	0.06 (.005)	0.05 (.002)	0.06 (.003)	0.02 (.002)

Source: Data from Judd and McQueen (1980).

Note: Values are means and (standard errors).

and Tracy 2008) by chasing and biting the intruders; and (4) Strecker (1928) reported a Texas tortoise with possible parental behavior toward its eggs, but herein lies a cautionary tale.

Strecker reported an observation of another person who had sent him three eggs of a Texas tortoise collected in Atascosa County. The eggs were supposedly laid on a "dry flinty ridge," and when the collector approached the tortoise, she "clumsily attempted to conceal the eggs, which were still warm." When a stick was thrust "toward the female she snapped at it viciously." The collector thought that the female was sitting on the eggs, one supposes as though it were a chicken. Whatever triggered the interpretation of this encounter is unknown, but we can definitely state that Texas tortoises do not defend their eggs, nest sites, or hatchlings, nor are they aggressive toward humans, nor do they incubate their eggs like a chicken!

Growth

Data on the growth of Texas tortoises of known age are available only for hatchlings and tortoises up to one year of age (Judd and McQueen 1980). The mean size of hatchlings and their instantaneous growth rates are given in table 5.1. Eggs hatched from August 27 to November 5. Tortoises grew from the time of hatching to early November, and then growth ceased until March. Tortoises grew throughout the spring and summer months, until the end of October, when the study concluded.

The posterior margin of the carapace of young tortoises exhibits considerable flexibility in response to muscle contractions. Apparently, during cold weather tortoise muscles are more tightly contracted than they are in warm weather; thus, individual tortoises often had smaller carapace length measurements during winter than they did in the preceding fall. The patterns of growth were similar for all measurements.

The greatest increase and most rapid growth were in weight, followed in order by carapace length, plastron length, carapace width, and shell height. Weight and carapace length showed a significant positive correlation with each of the shell measurements (Judd and McQueen 1980).

Although data are lacking for Bolson tortoises, size at hatching may be similar in all species of *Gopherus*. Grant (1936) reported a carapace length of 44 mm (1.7 in) and a weight of 19.7 g (0.7 oz) for a four-day-old desert tortoise. Patterson and Brattstrom (1972) found that the carapace length of six desert tortoises at hatching ranged from 43.7 to 47.2 mm (1.5 to 1.7 in), with a mean of 45.6 mm (1.6 in). Jackson et al. (1976) reported that four hatchling desert tortoises had a mean carapace length of 47.8 mm (1.7 in). Arata (1958) reported a mean carapace length of 43.4 mm (1.5 in) and a mean width of 42.1 mm (1.5 in) for gopher tortoises six hours after they had hatched.

Auffenberg and Weaver (1969), in their estimate of growth rate for Texas tortoises, suggested that carapace length at hatching was approximately 50 mm (2.0 in), and they calculated that the mean increase in carapace length in the first year of life was 50.3%. These data would indicate a mean carapace length at one year of age of 75.2 mm (3.0 in). Judd and McQueen (1980) found that mean carapace length at hatching was about 42 mm (1.7 in) and at one year was approximately 66 mm (2.6 in), indicating a mean increase in carapace length in the first year of 56%, similar to Auffenberg and Weaver's (1969) estimate. Consequently, the latter's overestimate of size at one year of age was due to their overestimate of carapace length at hatching. The growing season for Texas tortoises in the Lower Rio Grande Valley of Texas extends from March to November, about 245 days.

Pope (1939) reported that individual desert tortoises increased in carapace length from 46.9 to 70.4 mm (1.8 to 2.8 in) (50%) in one year. Conversely, Patterson and Brattstrom (1972) showed an annual increase in carapace length of 11.97% for desert tortoises and 12.50% for gopher tortoises. Their tortoises ranged from 50 to 74 mm (2.0 to 2.9 in) in carapace length and were probably in their first year of life. Thus, Texas tortoises may have a faster first-year growth rate than desert or gopher tortoises.

Auffenberg and Weaver (1969) estimated the growth of Texas tortoises older than one year using data on carapace length and growth rings on the scutes to identify size and age classes. Tortoises with just one growth ring ranged from 70 to 81 mm (2.8 to 3.2 in) in carapace length and were thought to be one year old. Tortoises were judged to reach sexual maturity at three to five years of age, with carapace lengths of 105 to 128 mm (4.1 to 5.0 in). Auffenberg and Weaver stated that growth slows markedly at about six years of age, at a carapace length of 130 mm (5.1 in), and stays at about 5% per year thereafter. They suggested that females grow faster than males and gave an absolute growth rate for males of 7.9 mm (0.3 in) per year and 11.1 mm (0.4 in) per year for females. This difference certainly cannot continue throughout life, because adult males attain significantly larger sizes than adult females (Rose and Judd 1982; Judd and Rose 1983).

Aufffenberg and Weaver (1969) showed that tortoise populations of closely situated but isolated lomas had differences in mean carapace length, which they concluded were due to differences in tortoise growth rates. Judd and Rose (1983) confirmed that there were marked differences in mean carapace length of closely situated but allopatric populations, but these differences were in the mean size of males.

Hellgren et al. (2000) reported that growth of juveniles in a population from Dimmit and La Salle Counties was 15.9 mm (0.6 in) per year, and they used this information to support their contention that tortoises could reach sexual maturity in five years. They provided data allowing the calculation of instantaneous growth in the carapace length of females between the ages of 4 and 15 years (age was estimated from growth rings of the scutes) of 3.5 mm (0.1 in) per year. Kazmaier et al. (2001b) provided growth curves for this same population that permit calculation of carapace length from hatching to ages from 1 through 16 years for females and males. They showed that growth in carapace length slows markedly after 8 years. Growth from hatching to 8 years of age is about 11.9 mm (0.47 in) per year for females and 12.5 mm (0.49 in) per year for males. From 8 to 16 years of age, growth in carapace length is about 0.6 mm (0.02 in) per year for females and 1.9 mm (0.07 in) per year for males.

The Silver Spoon

The Silver Spoon Effect (Grafen 1988) posits that individuals with high energy or an abundance of food in the initial stages of life will show accelerated growth rate trajectories not experienced by those lacking such resources. Larger size translates into the ability to consume a greater number of larger items earlier, giving larger individuals an environmentally supported advantage. If size and physical health are involved in the onset of sexual maturity, then larger individuals might also have an earlier reproductive advantage. Whether Texas tortoises experience the Silver Spoon Effect in the wild is not known; the underlying principle is reasonable but is probably muted because of the relative stability of the tortoises' available food and the vacillating weather patterns of drought and cold that affect their growth. In captivity, the obvious can be confirmed: hatchlings reared indoors at high temperatures with access to abundant food continue to grow throughout the winter, while those remaining in external enclosures do not. The individuals reared indoors attain sexual maturity (as evidenced by combat and courtship behavior) in three to four years, and one female laid a single egg in her fourth year. In addition, they can exceed the average length and weight of adults in the wild in six years.

Size and Age at Sexual Maturity

The sexes are indistinguishable at hatching, and males gradually develop dimorphic characters that are lacking in females as they approach sexual maturity. Auffenberg and Weaver (1969) stated that secondary sex characteristics are evident only in individuals with carapaces longer than 105 mm (4.1 in). However, Judd and Rose (1983) reported that the smallest individual identified as a female was 134 mm (5.3 in) in carapace length, and the smallest individual identified as a male was 138 mm (5.4 in). The smallest female x-rayed with eggs was 143 mm (5.6 in) in carapace length (Judd and Rose 1989). Hellgren et al. (2000) concluded that females in Dimmit and La Salle Counties reached sexual maturity at a carapace length of 131 mm (5.2 in), between four and eight years of age.

The sexual maturity of females can be unambiguously established by using x-rays to determine the presence of shelled eggs (see fig. 1.5). Thus, it is clear that some females in the Yturria and Reed Ranch populations in Cameron County, Texas, became sexually mature at a carapace length of 143 mm (5.6 in). Establishing the sexual maturity of males is more difficult; techniques that have been used on *Gopherus* species include observation of the development of secondary sexual characteristics (McRae et al. 1981; Mushinsky et al. 1994), observation of courtship behavior, and observation of sperm in seminiferous tubules (Germano 1994). Auffenberg and Weaver (1969) observed the beginning of the development of secondary sexual characteristics when they adopted a carapace length of 105 mm (4.1 in) as the minimum size at attainment of sexual maturity in the Texas tortoise. However, sexually defining characteristics are not completely established at this size and will not be so until the tortoises are considerably larger. We have seen elements of courtship behavior in captive males in their third year, with carapaces as short as 90 mm (3.5 in). Figure 5.4 depicts the general size distribution of a sample of males, females, and juveniles from the Yturria Ranch.

Sexual Dimorphism and Sexual Selection

Males develop longer and broader gular extensions on the plastron than do females, and they have a distinct depression in the inguinal region of the plastron (see fig. 1.7). In males, the xiphiplastron bones are thick and in older males appear as though they are rolled, but in females they are flat and thin. Males have longer tails than females and during the breeding season their subdentary glands are markedly larger (plate 3). Secondary sexual characteristics are not consistently apparent in males with carapaces shorter than 125 mm (4.9 in) (Judd and Rose 1989).

If females are more apt to select larger males as inseminating mates, and if larger, more aggressive tortoises are apt to dominate the reproductive scene,

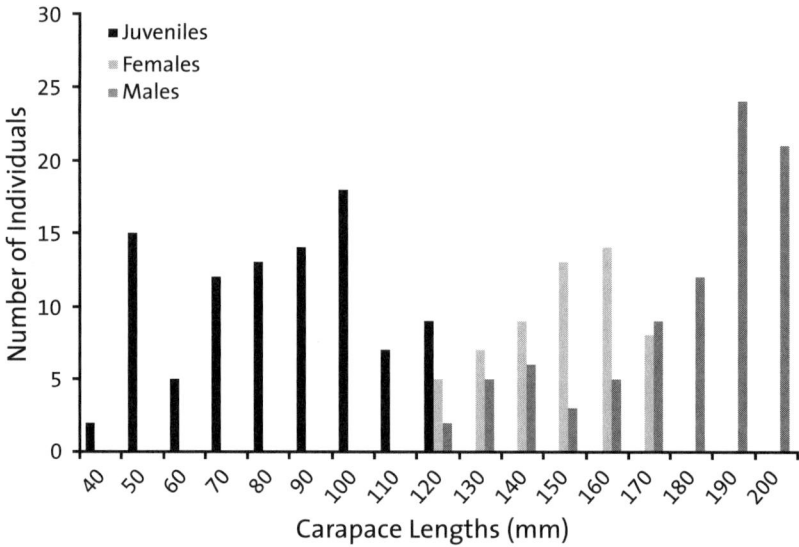

Figure 5.4. Carapace lengths at first capture of juvenile, female, and male Texas tortoises on the Yturria Ranch, Cameron County Texas.

and if we assume that size has a genetic component, we have a window to visualize how sexual size dimorphism may have arisen and how it is maintained in this small tortoise. This window, however, does not provide much insight into the mechanisms of the extreme sexual dimorphism seen in structural differences. It is important here to emphasize that discussions of sexual dimorphism take two paths, *sexual size dimorphism* and *sexual morphological dimorphism*.

Within the five species of *Gopherus*, it is difficult to envision scenarios that could account for the varying degrees of geographically related sexual dimorphism. Males of all five species are aggressive regarding reproduction, so the degree of sexual dimorphism cannot easily be attributed to male-male interactions. While male Texas tortoises attain a larger size than females, and the sexual size differences in other species are muted, the disparity in the Texas tortoise is geographical, with the greatest difference occurring in more coastal populations and becoming less pronounced farther inland. The structural differences between the sexes are still there, but the sizes attained by males and females in inland areas are virtually equal.

Bone is composed primarily of calcium salts, and adult females need lots of calcium for eggshell production. This might account for the thinner bones of females. Males do not have to sequester extra calcium for egg production and can therefore shunt calcium resources into developing and maintaining a more substantial carapace and plastron. This begs the question, however, of what is gained by developing structural differences, so we need to look at the known function of the dimorphic elements. Gular extensions are longer and thicker

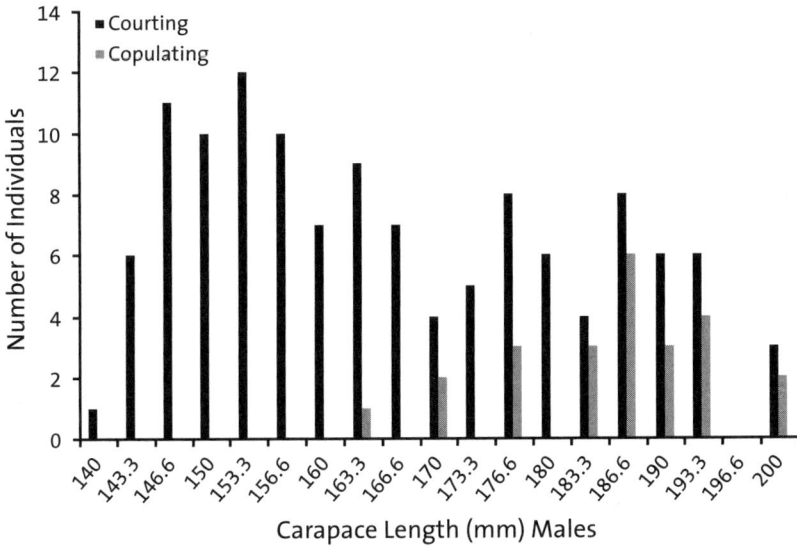

Figure 5.5. Size distribution of courting male Texas tortoises and those known to have had intromission, indicating that while males of various sizes court fervently, the larger individuals are more apt to mate.

in males; they are used in combat for ramming and flipping over rival males. In addition, ramming is used during courtship. The gular extensions of males are balanced between being long enough to serve as functional ramming tools but not so long that they interfere with feeding. In order for some larger males to open their mouth, they must reach their head over a platform created by the gular projection. This arrangement forces the tortoise to feed only on items within the plane of movement of the neck; access to items on the ground immediately in front of it is limited because the head does not extend far enough forward past the end of the gular extension to be able to tilt downward to where the mandible can swing open.

The male's extreme xiphiplastral concavity facilitates coupling in that the lower hind limbs of the male can be vertical while his plastron cups the carapace dome of the female. Bear in mind that this is a precarious position for the male while he grunts, slobbers, and jumps up and down on his two rear feet, matching the female's pivots one way and then the other. Male tortoises with the concavity filled with plaster were unable to mount and maintain position.

Female Texas tortoises do not normally copulate with a male tortoise that is smaller than they are. As in humans, the attainment of secondary sexual characteristics and the capacity to fertilize does not necessarily equal access to a mate. Moon et al. (2006) reported that gopher tortoises in Florida were polygamous and that larger males fertilized the majority of clutches. We report here (fig. 5.5) that out of 124 observations of courtship, there were 24 (19%) confirmed instances of copulatory insertion by a male. The smallest male that copulated had a carapace length of 163 mm (6.4 in), but 75% of the successful copulations were of males with carapaces longer than 175 mm (6.9 in).

The Texas tortoise is the only member of the genus *Gopherus* that does not construct a burrow, nor is it structurally modified to do so. It is higher domed than the other species, and its forelimbs are not broadly structured for digging. While one might muse that a tortoise has plenty of time to spend digging, the construction of burrows even in soft soils demands energy—energy that could be used for other facets of maintenance, reproduction, and survival. Tunnels of gopher tortoises are legendary; up to 30 feet in length and 12 feet deep at the inner terminus. The question therefore arises, what adaptive value do tunnels have? They provide protection from predators and from environmental extremes, such as heat and cold. So what factors could be responsible for the lack of burrowing behavior in the smallest tortoise, and ostensibly the one most vulnerable to the extremes of climate and predators? Applying Occam's razor suggests that burrow construction is a primitive characteristic because it occurs in four of the five extant species that occur in two closely related lineages. Somewhere along the line after geographical separation, the Texas tortoise lost the behavior of constructing extensive burrows but retained the behavior of constructing pallets, that is, burrow entrances. We suggest that the lineage that was to become the Texas tortoise was isolated along the coastline, where it was associated with warmer temperatures and unstable soils, and its current distribution in Texas is the result of a northward and westward movement from the coast. With its smaller size, it was better able to take advantage of other animal burrows and refuges. The larger size of males may be a reflection of their more nutritious diet in the coastal areas, which favored their growth, while females had to convert much of their sequestered energy to reproduction. The big males in coastal areas that emphasize the sexual size dimorphism may also live longer than males in inland areas, but this awaits documentation.

Population densities, sex ratios, and territorial defense might also account for sexual dimorphism in that at low densities, females are a resource worth fighting for, and at sex ratios near equality, energy expended in agonistic behavior takes away from energy spent reproducing. There is no evidence that Texas tortoise males are territorial, that is, defend a given space. In chapter 8 we present evidence that females copulate more frequently with larger males. Sexual dimorphism is a fascinating and fruitful area for future research and we expect that to elucidate this important aspect, the food quality of known diets in different geographical areas, as well as the energy allocations to growth, maintenance, and reproduction, will have to be ascertained.

Mating System

Classification of mating systems typically reflects the degree that males and females bond during the breeding season. In this arrangement, monogamy is the association of one male with one female, with exclusive mating between the pair. Polygamy includes all forms of multiple mating. Polygyny is a subset of polygamy in which one male associates with two or more females at a time. Polyandry is an association between two or more males and one female at a time. In promiscuity there is no prolonged association between male and female and multiple matings occur by at least one sex.

Emlen and Oring (1977) developed the following ecological classification of mating systems based on the ability of one sex to monopolize or accumulate mates. Although developed for birds, this classification system has been applied to other animal groups, including tortoises of the genus *Gopherus*. Polygyny is defined as when "individual males frequently control or gain access to multiple females." Polygyny is divided into three subcategories: (1) resource defense polygyny, in which males control access to females indirectly by monopolizing critical resources; (2) female (or harem) defense polygyny, in which males control access to females directly, usually by virtue of female gregariousness; and (3) male dominance polygyny, in which mates or critical resources are not economically monopolized. Males aggregate during the breeding season and females select mates from these aggregations. Because male gopher tortoises engage in combat that leads to access to females, Douglas (1986) inferred that this species has a female defense polygynous mating system. Conversely, Johnson et al. (2009) suggested that other behavioral characteristics of *Gopherus* species are consistent with the mating system identified as scramble competition polygyny by Wells (1977) and Schwagmeyer and Woontner (1986). Scramble competition polygyny features intrasexual selection based primarily on the results of competitive searching for mates instead of combat (Schwagmeyer and Woontner 1986). Moon et al. (2006) reported that gopher tortoises have a promiscuous mating system, with larger males fertilizing the majority of clutches. Clutches produced by larger females tend to be sired by a single male, and clutches of smaller females tend to be sired by multiple males. They did not attempt to classify the mating system. Johnson et al. (2009) used telemetry to distinguish between female defense polygyny and scramble competition polygyny in gopher tortoises. They found that the patterns of dispersion and burrow use support scramble competition polygyny in this species. Johnson et al. (2009) seemingly argued that behavioral and morphological variables suggest that gopher and Bolson tortoises have scramble competition polygynous mating systems and that desert and Texas tortoises have female defense polygynous systems. Unfortunately, the last sentence of their paper indicates that all four species have female defense polygynous mating systems, despite the title of their paper stating that gopher tortoises have a scramble competition polygynous system.

Data from the field and from captivity show that both sexes of the Texas tortoise are promiscuous. Both sexes have home ranges (Rose and Judd 1975; Judd and Rose 1983). General activity is concentrated in the central area of the home range, with less probability of its occurrence per unit area toward the periphery (Judd and Rose 1983). Thus Texas tortoises do not patrol or defend a perimeter. Likewise, they do not defend home burrows because they do not construct them themselves, and pallets are used on an opportunistic basis (Rose and Judd 1982). However, males do engage in combat, possibly for access to females (Judd and Rose 1983). Males often trail females for several days during the breeding season, and we have noted (Judd and Rose 1983) that if selection favored males mating with as many females as possible, this would account for the greater vagility (freedom to move about) of males. We also found evidence for polygynous mating in our observation of the same male trailing and courting different females within a 24-hour period (Judd and Rose 1983).

When we tested the dispersal of activity centers on a study grid in Cameron County for randomness, we found that the activity centers were highly clumped and seemed to be closely associated with the distribution of prickly pear, that is, there was a high frequency of tortoise activity centers in quadrants having relatively large clumps of prickly pear (Rose and Judd 1975). In fact, 62% of the tortoises observed on the grid were under or in the immediate vicinity of prickly pear. We did not test the dispersion of the sexes separately. With the mean minimum polygon home range sizes we reported for males (0.47 ha [1.2 ac]) and females (0.29 ha [0.7 ac]) (Judd and Rose 1983), 7 males and 11 females could be fitted onto the 3.3 ha (8.2 ac) grid without overlap of home range within a gender. The smallest number of adults occurring on the grid in a given year over a five-year period was 15 males and 15 females (Judd and Rose 1983). Clearly, there is considerable overlap of home ranges within each sex, but much greater overlap among males. Given the small size of the tortoises relative to the size of their home range, encounters between tortoises would likely be rare even though there is overlap among home ranges, unless males actively search for females.

Texas tortoises exhibit a clumped dispersion pattern in relation to critical resources such as prickly pear, which provides food, water, shade, and protection from predators. There is no evidence that members of either sex defend such resources or any area of their home range (that is, they are not territorial). Males actively search for females during the reproductive season and engage in combat with other males for access to females. It is likely that males are sometimes able to mate without encountering another male. Once a female is located, the male initiates courtship. If the female is not receptive, the male remains in the immediate vicinity and engages in combat with other males that may approach the female. A given male will remain with a female and court her repeatedly until he is successful in accomplishing intromission and sperm transfer, until he is displaced by another male, or until his persistence is exhausted. Once a male mates or moves on, he likely concentrates on searching for a different female, for his potential fitness is increased the more females he mates with.

The length of time a male spends defending a given female is not known, but our observations suggest that it is not more than a few days. Nevertheless, the mating system is best characterized as female defense polygyny. The cues males use to locate females are of great interest. Vision no doubt plays a role (Aufffenberg and Weaver 1969; Weaver 1970), but likely primarily over short distances. Olfaction is more likely over longer distances. Secretions from subdentary glands (Auffenberg 1969; Rose et al. 1969; Rose 1970) wiped on vegetation may enable males to distinguish between other males and females. Cloacal secretions in urinary deposits might also help identify females.

Color and Patterning

Auffenberg and Weaver (1969) described the color and patterning of hatchlings in considerable detail. Vertebral and costal scutes of the carapace have creamy white to yellow centers surrounded by dark brown to black areas (plate 24). Anterior marginals have creamy white to yellow areas limited to the posterior edges of each scute, lateral and posterior marginals have light pigments restricted to the free edges of the scutes, and the remaining areas of the marginals are dark brown to black. On the plastron, dark pigment is limited to seams between the scutes, except on femoral scutes, which are all dark except for the free edges. The throat, the subdentary area, and the medial surfaces of the limbs are whitish to light gray, except for scales at the base of the limbs, which are dark. Proximal elements of the limbs are dark. The top and sides of the head are light except for a yellowish stripe on the side of the head that runs from the anterior corner of the eye to the angle of the jaw. As tortoises grow (get older), pigmentation changes from black and yellow to brown and yellow to a uniform brown or "horn" color (plate 25).

The description above is best seen as a general model because there is considerable individual variation in the pigmentation of young tortoises. On rare occasions, the carapace of a hatchling tortoise is brownish black, with only the scute centers a dull yellow. As these tortoises age, they maintain the dark pigmentation; their heads are noticeably large, and they are invariably males. They are aggressive toward other males, have accelerated growth, and begin courting at an early age. To our knowledge, these individuals are known only from captivity (see plate 28).

The other extreme is a brightly colored hatchling, with stunning yellows and a bright tan. The skin may be orange, and these individuals develop into the light tan "blonds." We have never observed these two dark and light color extremes within the same clutch.

Geographic Variation in Size

Auffenberg and Weaver (1969) reported differences in mean carapace length of tortoises on closely situated but isolated lomas. They concluded that the size differences were due to differential growth rates on different lomas. They did not include values that would enable calculation of the variance of their samples. Consequently, the significance of the observed differences in the mean values cannot be tested. Means for two of the populations they studied were based on small numbers of tortoises (Port Isabel Loma: 7 males, 1 female; Mesa del Gavilan: 5 males, 10 females) and may not represent the actual population size and structure on those lomas.

We compared the sizes of tortoises from 4 km (2.5 mi) south of Hargill, Hidalgo County, and the Yturria Ranch, Cameron County, with data provided by Auffenberg and Weaver (1969) for Loma Tio Alejos, Mesa del Gavilan, and Port Isabel Loma (all in Cameron County) (Rose and Judd 1982). We found that males of the Yturria Ranch and Hargill populations were significantly larger than females of the same populations and that Yturria Ranch males were significantly larger than Hargill males. Conversely, there was no significant difference in the size of Yturria Ranch and Hargill females.

The significance of the differences in tortoise sizes that we observed compared with data provided by Auffenberg and Weaver (1969) could not be assessed, but mean sizes of the Loma Tio Alejos males and the Hargill males were similar. Because Yturria Ranch males were significantly larger than Hargill males, it appears likely that Yturria Ranch males are also significantly larger than Loma Tio Alejos males. Mean sizes of males of Mesa del Gavilan and Port Isabel Loma were markedly larger than the mean for Yturria Ranch males, which suggests that Mesa del Gavilan and Port Isabel Loma males may be significantly larger than Loma Tio Alejos males. Females exhibited far less variation in size than males. These data support Auffenberg and Weaver's assertion of marked differences in mean carapace length in closely situated but allopatric populations (at least for males).

The largest female reported by Hellgren et al. (2000) measured 165.9 mm (6.5 in) in carapace length. However, we showed that 17 of 29 gravid females in two Cameron County populations were larger (Judd and Rose 1989).

Some authors prefer to use plastron length as the chief measure of size, but the dimorphic plastrons of Texas tortoises hamper male-female comparisons. There are two ways to measure tortoise carapace length. The straight-line method is the distance from the anterior nuchal or the *V* formed from its absence to a vertical line that marks the posterior edge of the carapace. The inline method uses a flexible tape that is extended from the nuchal along the centrals and terminates at the posterior edge of the large last marginal.

Auffenberg and Weaver (1969) suggested that variation in mean size among populations is probably due to differences in food and its availability. We stated that it is difficult to imagine how a food-related growth response could be expressed in males but not in females (Rose and Judd 1982). However, we noted that prickly pear was markedly more abundant at the Yturria Ranch than it was at Hargill, which could possibly explain the larger mean size of the Yturria Ranch males.

Longevity and Senescence

Tortoises are often cited as examples of long-lived species, but in fact, little is known about their longevity. Judd and McQueen (1982) reported on the longevity of captive Texas tortoises. McQueen had twelve tortoises (six males, six females) that he captured in the field as adults in 1955, and he also had three males and one female that he inherited from A. H. George of La Villa, Texas, that had been in captivity since 1927. When data were recorded on all of these captive tortoises in 1979, one group was more than 24 years old, and the other group was more than 52 years old. Fertile eggs were laid by five of the six females that were more than 24 years old in 1979. The female that was more than 52 years old did not lay fertile eggs in 1979 or 1980. All three of the males that were more than 52 years old courted females in 1979 and 1980, and all but one of the males more than 24 years old courted females. If it takes tortoises at least six years to reach adult size, the tortoises that were more than 24 years old were at least 30 years old, and females were clearly not reproductively senescent at that age. If the same reasoning is applied to the female that was more than 52 years old, that means she was at least 58 years old. Data are far too meager to suggest that females ever reach reproductive senescence.

Survivorship

Understanding the survivorship of such a long-lived species is critical in conservation planning. Intuitively, one expects males of this species to have higher survivorship than females, and juvenile survivorship to be less than that of females. After all, females undergo the extra rigors of egg production and nest construction, both of which increase their physiological stress and vulnerability to predation. In addition, the thinner shell bones of females are more easily broken. Juveniles are subject to a series of mesopredators that do not negatively impact adults, such as snakes, wood rats, and birds, as well as those that do prey on adults.

Two studies estimated survivorship of the Texas tortoise. Hellgren et al. (2000) evaluated survivorship in a low-density inland population in Dimmit and La Salle Counties, Texas, using the recapture of telemetered tortoises and a regression analysis of frequency on age as determined by scute annuli. They estimated an endpoint survivorship of 0.774 (77.4%) for adult females and 0.828 for adult males. They also suggested that only 5 out of every 1,000 hatchling

Table 5.2. Annual survivorship estimates for Texas tortoises on the Yturria and Reed Ranches, Cameron County, Texas

Age/sex category	Yturria Ranch			Reed Ranch		
	Estimate	SE	Confidence limits	Estimate	SE	Confidence limits
All tortoises	0.913	0.012	0.886–0.933	0.813	0.028	0.753–0.861
Males	0.939	0.013	0.908–0.960	0.860	0.031	0.787–0.910
Females	0.925	0.020	0.877–0.956	0.754	0.054	0.634–0.844
Juveniles	0.687	0.061	0.557–0.794	0.424	0.221	0.111–0.812

Source: Data from Rose et al. (2011).

tortoises survive to age 20. Rose et al. (2011) evaluated recapture data generated from the Yturria Ranch (1972–87) and Reed Ranch (1977–87) populations in Cameron County, Texas. The timing of known deaths was distinctly different for these two sites, with 9 tortoises (3.79%) found dead on the Yturria Ranch from 1975 to 1987, and 16 known tortoise deaths (10.59%) on the Reed Ranch from 1977 to 1979. Annual apparent survival estimates by sex and age group indicate that survivorship was higher for both males and females on the Yturria Ranch than on the Reed Ranch, although it was significantly higher only for females (table 5.2). Yturria Ranch males showed substantially higher survivorship, although not significantly so. Data indicated that annual survivorship estimates might have been greater for juveniles on the Yturria Ranch, but further studies are needed to clarify this issue. We point out that the two ranches are only 6.4 km (3.9 mi) apart and emphasize that only long-term simultaneous studies at multiple sites can clarify this important aspect of the life history of the Texas tortoise.

6

Sensory Modalities

> In a time of drastic change it is the learners who inherit the future. The learned usually find themselves equipped to live in a world that no longer exists.
>
> —Eric Hoffer,
> *Reflections on the Human Condition*

Although our knowledge of the sense organs of the Texas tortoise is limited, the description in this chapter of the tortoise's senses of olfaction, vision, touch, hearing, and taste may provide some insight into how it perceives its world.

Olfaction (Smell)

Little is known about the olfactory acuity of Texas tortoises, but they behave as though they smell objects and individuals. They frequently push their nose into a food item prior to its consumption or rejection. Generalized head bobbing seems to be in response to olfactory cues associated with food recognition (McCutcheon 1943; Eglis 1962), and Auffenberg (1969) reported that the more elaborate head bobbing sequences of courtship and combat were derived from those associated with recognition of food. In an unpublished series of experiments, we discovered that dehydrated Texas tortoises did not find or associate with pond water that was covered with cheesecloth, whereas box turtles (*Terrapene ornata*) and plains leopard frogs (*Lithobates blairi* or *Rana blairi*) readily found water in the dark. These experiments suggest that olfaction is not involved in tortoises' detection of water outside their visual field. However, in another experiment, Texas tortoises were able to find and access food that they could not see that was housed in a metal bowl. One tortoise learned to use its head to tip the bowl by positioning its chin on the rim and pulling the bowl over.

From a series of experiments, Weaver (1970) reported that females distinguished males from other females by cloacal scent but that males could not distinguish females by cloacal scent, morphology, or movements. He thus concluded that males might rely more heavily on subdentary gland secretions to distinguish between the sexes.

General observations of wild and captive Texas tortoises reinforce our view that olfactory acuity is a moderately developed sense. There appeared to be no diminution of courtship activity by males or females in captivity when the nares were plugged with wax. Because of the inconsistent degree of head bobbing, Weaver (1970) stated that "head bobbing in *berlandieri*, and probably in the other *Gopherus* species, is primarily a sniffing movement."

Galeotti et al. (2007) reported that both sexes of the Hermann tortoise (*Testudo hermanni*) discriminated between conspecific odors and odors from other species, but only males distinguished sex and sexual maturity by smell. Such a well-designed study using Texas tortoises as model organisms would be helpful in elucidating the role of olfactory communication in North American tortoises.

Secretions from the subdentary glands of Texas tortoises contain a mixture of fatty acids, among other components (Rose 1970). Both male and female Texas tortoises responded positively to a mixture of these fatty acids painted onto a tortoise model positioned along a wall of an enclosure. There was no difference in the responses of males and females to the control, but 59% of the tortoises gave positive responses to the model painted with the solution. All of the responding males pushed the model and six turned it over during their encounter. No male attempted to mount or bite the model, indicating that their response was agonistic and not related to courtship (Rose 1970). These experiments were carried out at high population densities during the peak reproductive period, when courtship and combat were vigorous. How subdentary gland secretions influence behavior under normal circumstances has not been addressed. The role of olfaction in homing behavior is treated in chapter 8.

Vision

The eyes of Texas tortoises are typical of those of other reptiles, with well-developed lids. The irises of Texas tortoises are a uniform brown (plate 3). Lacrimal and Harderian glands keep the eyes moist, but the lacrimal glands do not have drainage tubes to the pharynx, which causes the eyes to be excessively moist at times. These are also the glands that secrete excess salt in sea turtles and crocodiles, but we have seen no evidence (salt crystals around the eyes or nose) that the glands in Texas tortoises function in this manner. However, salt-loading experiments might confirm this salt excretion function, because some Texas tortoises occur on barrier islands and near mainland beaches.

Texas tortoises inhabit environments where visual acuity may be more highly adaptive than olfaction. Individuals are known to have instigated escape behavior (ceased feeding and moved rapidly to cover) when an observer was over 40 yards away. In addition, some of our captive individuals readily move toward a feeding station when they observe a person approaching from as far away as 40 yards; others bolt to shelter. For some unknown reason, Texas tortoises perceive humans as threats, and when they see a human, most will resort to escape behavior. This behavior is not noted when a large mammal such as a deer or dog approaches. How tortoises differentiate among potential predators is not known. However, some Texas tortoises, even some hatched in captivity, continue their escape behavior on seeing a human even after many years of daily contact and feeding.

Several authors (Grant 1960; Auffenberg 1969; Auffenberg and Weaver 1969) have noted that Texas tortoises selected red foods, such as mature prickly pear tunas, over food of other colors. No doubt, when mature red fruits are available, they are readily consumed, as evidenced by remnants on jaws, pink urine, and seeds in fecal pellets. However, when we tested captive tortoises in an outside enclosure and presented them simultaneously with food colored red (tomatoes and prickly pear tunas), purple (eggplant), yellow (squash), and green (romaine lettuce, prickly pear cladophylls, zucchini), romaine lettuce was selected first and green cladophylls last. Perhaps confinement altered the basis of preferences from color to water content. We point out that there are many areas in South Texas where prickly pear is scarce, and tortoises would require minimal color differentiation of food items. Therefore, color vision in the Texas tortoise appears to be alive and well, but the degree of its influence on behavior remains unexplored.

Depth perception is well developed in Texas tortoises soon after hatching. Individuals left on solid tabletops do not walk willy-nilly off into space as aquatic turtles do. We have observed them making a "conscious evaluation" of the cliff, and if it is low enough, they feel forward with their forelimbs and then slide forward and down, the same as box turtles do. Apparently good depth perception is an asset to terrestrial turtles but is not a critical factor for aquatic ones. If a Texas tortoise is placed on a table with a clear glass top, it will panic and scramble backward—to its detriment, as it cannot see the rear edge of the table. When Texas tortoises are held they frequently thrash their legs frantically, most notably their forelimbs, and appear to be trying to back up. This response can undoubtedly be attributed to the visual void that presents itself to them as they are being held. It is easy to hold two tortoises at once, one in each hand, but we advise against attempting to hold three! Patterson (1971c) demonstrated depth perception in Texas and desert tortoises, as well as in box turtles, using a visual cliff.

Touch and Hearing

Little is known about the physiological aspects of touch and hearing in the Texas tortoise. Although the nonshelled areas are covered with thick skin and bony plates, the limbs are withdrawn reflexively when touched, as is the head when the area around the nasal openings is touched. Tortoises respond immediately to heat but less vigorously to cold.

The hearing apparatus is difficult to see externally, as the surface is a rather thick membrane (see plate 3) that connects internally to the middle ear via a small bone, the columella. Earlier authors (Gulick and Zwick 1966; Patterson 1966; Campbell and Evans 1967) showed that turtles and tortoises have sensitivity to sounds below 1000 cycles per second (cps), and the optimal range may be between 200 and 500 cps. Although the possibility is unstudied, the inner ear of Texas tortoises may be able to perceive ground vibrations, because when tortoises are resting, their extended head often rests directly on the ground.

Various chelonians from numerous families are known to make vocalizations (Galeotti et al. 2005b), usually associated with courtship. While vocalization has not been extensively studied in tortoises, and its role in communication is not well understood, there are some peculiar examples. The South American red-footed tortoise (*Geochelone carbonaria*) makes clucking sounds generally associated with courtship, but juveniles may make the noise as they forage (Campbell and Evans 1972). Travancore tortoises (*Indotestudo elongata travancorica*) from southern India may call in a chorus, usually at night, producing the most spectacular example of chelonian vocalization. The function of the chorusing in these tortoises is unknown, but it was suggested, because the tortoises eat frogs, to be a mechanism to entice food (Auffenberg, pers. comm.; Campbell and Evans 1972). This is a splendid example of one of nature's Darwinian traps: showing up for sex only to be eaten by a tortoise.

Male Texas tortoises may make hissing noises during intense courtship. There is no evidence that they hiss prior to mounting, although during rigorous courtship, pursuit, and combat they may emit a low-frequency hissing that we interpret as simply forceful exhalations. Attempts to establish courtship-related vocal behavior in Texas tortoises by Auffenberg (pers. comm.) were unsuccessful. Courting male Hermann tortoises (*Testudo hermanni*), on the other hand, emit fast, high-pitched calls that are part of a multiple-signal system used by females to select higher-quality mates (Galeotti et al. 2005a, 2005b).

Taste

The large tongue of Texas tortoises has deep clefts and is involved with swallowing. It is also used to reject food judged to be distasteful. With this observation, we assume that taste is the final arbiter among the primary senses as to what is to be swallowed.

Texas tortoises express preferences for certain foods when presented with options. Those that eat white cauliflower will also eat the green version; those rejecting white also reject the green (plate 26). The green leaves of bok choy are usually eaten, but the white stems may be rejected. Turnips, which are the same species as bok choy, are rejected, as are any vegetables with mustard oils such as mustard greens and radishes. Cactus tunas coated with salt brine or alum are rejected, as are tunas injected with vinegar. Most tortoises consume cantaloupes and watermelons, choosing the yellow and red centers over the rinds. Texas tortoises readily eat commercial tortoise chow (see chapter 11), if it is softened slightly with water, and presumed sniffing precedes the first bite. Herein lies the rub; it is impossible to separate olfaction, taste, and the interplay of visual cues that are enmeshed in an individual tortoise's feeding repertoire. We suggest that the sensory modalities of tortoises are in need of clarification, and the use of nasal plugs and anosmic agents might be a helpful start.

7

Temperature Regulation, Water Use and Retention, and Photoperiod Response

In measuring the ecological, anatomical, physiological, or behavioral parameters of any organism, we can never be sure our data depict its normal activities. The act of detecting, measuring, and describing such parameters, indeed, our very presence, increases the degree of uncertainty and is likely to affect the outcome.

—Thomas Simpson, extrapolation of the
Heisenberg uncertainty principle

The challenges posed by temperature regulation, water use and retention, and photoperiod responses are important to all organisms, but particularly to those such as the Texas tortoise that live in arid environments. This chapter will explore how the Texas tortoise meets these challenges.

Temperature Regulation

Tortoises belong to that broad group of vertebrates that cannot effectively maintain a metabolically optimum body temperature over a wide array of environmental temperatures. These poikilothermic creatures are not, however, at the mercy of their environment. Texas tortoises are able to inhabit potentially lethal environments through various behavioral adaptations, and they use rather unsophisticated physiological techniques when they encounter thermal excess.

Temperatures in the Texas tortoise's environment can easily reach lethal extremes throughout much of the year, yet it is the only member of the genus *Gopherus* that does not dig a substantial burrow that might mitigate negative thermal impacts. Burrow construction is no doubt energetically costly in any soil, but it would be much more so in compacted clays. However, in inland populations, such as those in Dimmitt County, Texas, where soils are sandy loam, pallets are rarely constructed. One can conclude that Texas tortoises do not dig burrows because they have adopted less energetically expensive alternatives. Their small size generally allows them to find shelter in the burrows of other animals, in wood rat middens, or simply under clumps of grass or cactus. Increased exposure to predators is a downside to not constructing burrows.

Texas tortoises generally curtail activities during the summer heat (Voigt and Johnson 1976), but even so, they may be active early and late in the day (Rose et al. 1988). Before exiting retreats in the morning, they generally spend a little time warming and are not usually active until their body temperature reaches 28°C (82.4°F). Deep core temperatures of 40°C (104°F) may be critical for exposed tortoises (Grant 1960) because the righting response is impaired, indicating neuromuscular compromises. The critical thermal maximum was determined in the laboratory to be 42°C (107.6°F) (Hutchison et al. 1966). Lowe et al. (1971) reported that Texas tortoises had a supercooling limit of -5.25°C (22.5°F) and a freezing point of -0.38°C (31.3°F). Tortoises at the northern limits of their geographic range may experience temperatures below -5.25°C on rare occasions, but minimal cover appears sufficient for tortoise survival.

In our study of the thermal biology of Texas tortoises, 91% of active tortoises had body temperatures between 30°C and 35°C (80°F–95°F), and we found a positive relationship between air temperature and the body temperature of active tortoises (Judd and Rose 1977). There was also a positive relationship between substrate and the body temperature of inactive tortoises. It follows, then, that the average body temperature in the spring (31°C [87.8°F]) is significantly lower than it is in summer (33°C [91.4°F]) and significantly higher than in the winter. The lowest recorded body temperature of an active adult tortoise was 24.1°C (75.4°F), and the highest was 39.8°C (103.6°F), 2.2°C below the critical thermal maximum. The tortoise's maintenance of relatively high body temperatures is an obvious digestive advantage for an herbivore.

Mean active body temperatures of the other *Gopherus* species (Rose 1983) are similar (30°C–35°C [86°F–95°F]). The finer points of difference are probably related to differences in body size and behavior associated with tunnel construction and use.

Texas tortoises heat faster than they cool (Voigt and Johnson 1977; Rose and Judd 1982), so they are not completely at the mercy of their thermal environment. When heat stressed, tortoises quicken their movements, start to salivate, and pant with their mouth open. Usually they defecate and urinate and begin to flick sand with their forelimbs. Frothing and salivating on the expanded and extended neck, and urinating on the inguinal skin, provide opportunities for evaporative heat loss, significantly delaying the attainment of a lethal body temperature (fig. 7.1). If no relief is obtained, first the hind limbs are paralyzed, then the forelimbs, leaving the tortoise unable to avoid its fate. Larger tortoises

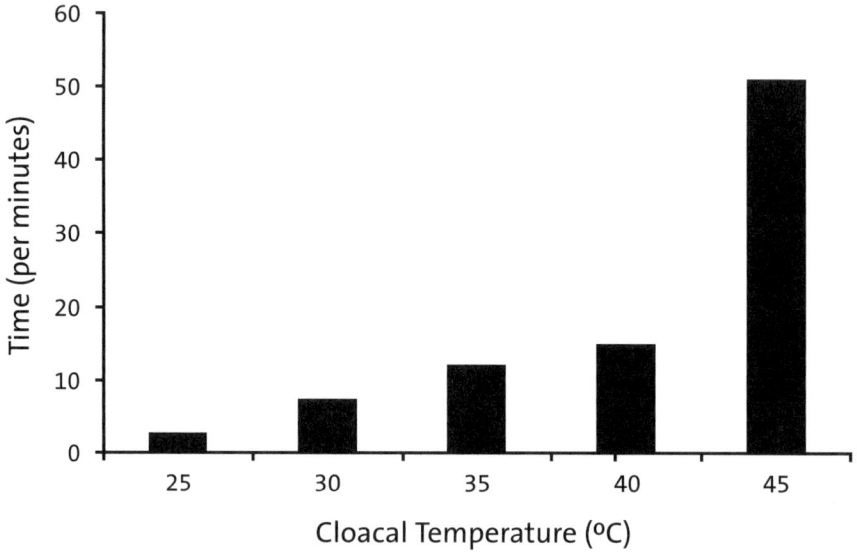

Figure 7.1. Time required for a 5°C increase in cloacal temperature of Texas tortoises at an ambient temperature of 55°C.

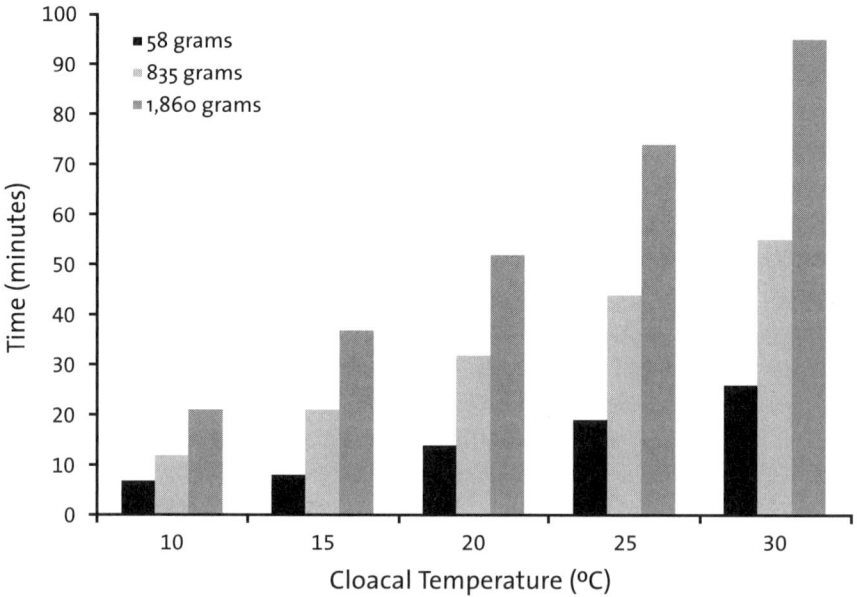

Figure 7.2. Time required for a 5°C increase in cloacal temperature of Texas tortoises of differing weights at an ambient temperature of 55°C.

are able to retard thermal stress to a greater degree than smaller ones (fig.7.2) because of their greater mass. We found two free ranging tortoises in thermal stress on the Yturria Ranch near noon, and both were juveniles.

It is unfortunate, but as environmental temperatures decrease around a poikilotherm, its metabolic functions also decrease; so, gradually, muscle activity is compromised, and the animal loses the ability to seek adequate shelter. The most notable injury observed in tortoises is hind limb paralysis. This is thought to be caused by the rear of the carapace not being adequately protected, which allows the underlying tissue, including the spinal cord, to freeze. There is no recovery of function.

Much has been written about whether turtles and tortoises heat faster than they cool (Bethea 1972; Voigt and Johnson 1977; Spray and May 1972; Voigt 1975; Weathers and White 1971). The disparity in results for the Texas tortoise (heats faster than it cools), the gopher tortoise (cools faster than it heats), and the desert tortoise (heating and cooling rates are equal) may be in part a reflection of small sample sizes and differences in treatments. Spray and May (1972) restricted movements of desert tortoises such that their limbs could not be retracted, and while Voigt and Johnson's (1977) animals were allowed to move about, they placed temperature probes inside the body through holes drilled in the carapace and plastron, at the base of the skull, in the middle of the neck, and in the hind leg below the knee. All of this work confirms that many reptiles have the ability to regulate blood flow when exposed to varying thermal regimes. This regulation is probably under control of the hypothalamus in the brain, which functions as the body's thermostat. Heating the heads of six Texas tortoises maintained at 22°C (71.6°F) with a concentrated delivery of warm (30°C [86°F]) air triggered a twofold increase in heart rate within one minute, while no increase was noted with air at 22°C. In addition, warm air delivered to the limbs or posterior carapace triggered no increase in heart rate.

A body of work suggests that sex is determined in most reptiles by the environmental temperature to which the eggs are subjected during a critical period of development. We know of one study that was initiated to evaluate temperature-dependent sex determination (TSD) in Texas tortoises by personnel at Gladys Porter Zoo in Brownsville, Texas, but the study was not completed. An investigation that compared TSD in Texas, Bolson, and desert tortoises would be a fertile area of research.

Water Use and Retention

Texas tortoises are physiologically adapted to a dry, harsh environment and are able to withstand serious droughts. Free water is, of course, available after rains but the tortoises obtain most of their water, called preformed water, from moisture in their food, cactus being the primary source when available. The red fruit, when ripe, provides the most readily available water source (85%)

that is not burdened with acid, as the green cladophylls are. Soft, newly formed cladophylls are higher in water content (94%–95%) than the more fibrous, older ones (91%–92%) that may be available year-round (Rose and Judd 1982). How tortoises metabolize the organic acids (pH 3.5–4.8) found in cladophylls is an open area for research. Perhaps, the high levels of ammonia found in the Texas tortoise (Baze and Horne 1970) are a response to organic acids in the diet. Oxalosis from eating mature cactus cladophylls, or vegetables high in oxalic acid, has not been reported in Texas tortoises.

Tortoises drink through their nose (external nares) and not through their mouth. After long periods without water, they may immerse their head in it completely, extending their neck, which can be seen expanding and contracting as the water is taken in. Captive tortoises seem not to recognize water as such in a structured basin but may drink readily from water out of a hose. If placed in a container partially filled with water, they might drink, and if they do, they usually defecate and evacuate their bladder, thereby making room for incoming water and also flushing the bladder of its concentrated urinary salts. Anecdotal accounts suggest that Texas tortoises may take water in through the cloaca when partially immersed, but to our knowledge no one has proven this experimentally. Small free-ranging tortoises require drinking water, but older ones may drink so infrequently that the word "never" comes to mind.

Because the skin and shell are impermeable to water, primary water loss is through breathing, defecation, and urination. And, because of the urinary bladder's ability to reclaim water by precipitating uric acid, little water is lost via that route. Tortoises are a system for water retention, not loss, and as arid land inhabitants they are extremely efficient at water retention.

Texas tortoises maintained in captivity were weighed to ascertain water loss under various environmental situations (Olson 1976, 1987, 1989). A tortoise weighing 120 g (4.2 oz) and one weighing 2150 g (75.8 oz) lost 0.78% and 0.03% of their weight per hour, respectively, during the summer. The larger tortoise lost water at a rate 26 times less than that of the smaller one. Tortoises maintained during winter (not allowed to burrow) at two temperature regimes (20°C–25°C and 5°C–10°C) lost 0.014% to 0.130% of their weight over 24 hours at the higher temperatures and 0.003% to 0.060% at the lower. Obviously, relative humidity will affect such experiments, but Olson's studies illustrate the need for carefully controlled experiments with sufficient numbers of tortoises of varying sizes to get a good handle on this important aspect of the physiology of Texas tortoises.

Photoperiod Responses

Many functions of animals vary with the time of day or year. Activity, for example, is restricted to the day (diurnal) or night (nocturnal) in many species, and reproduction is often confined to certain seasons. Regular, time-dependent variations in function are termed biological rhythms. Among the most obvious rhythms are daily variations in activity. Rhythms that cycle on a daily basis (24 hours) are termed circadian rhythms. Rhythms that continue in the absence of obvious environmental cues are termed endogenous, and rhythms that fail to persist in the absence of an environmental cue are termed exogenous. Even endogenous rhythms often must be occasionally exposed to an environmental cue to keep them from drifting out of phase with the environment. An environmental cue that is capable of setting the phase of a biological rhythm is termed a phasing factor, or zeitgeber. Of all the known zeitgebers, those most commonly successful in entraining endogenous circadian rhythms are daily light cycles or changes in light intensity. Sunrise is likely the natural cue (Zug 1993).

The role of photoperiod in regulating the circadian rhythm of activity in Texas tortoises has not been studied. Nevertheless, the tortoises are clearly diurnal, although temperatures at night in late spring, summer, and fall are sufficiently high to permit activity. The absence of nocturnal activity strongly suggests that the circadian rhythm of activity is determined by light. Because tortoises cease activity before sunset, light intensity or angle may serve as a cue to trigger the cessation of activity. However, Auffenberg and Iverson (1979) stated that the daily activity cycle of terrestrial turtles is in large part a response to temperature and moisture conditions rather than to light (although light is important insofar as it must be present).

Rhythms may be seasonal or annual. The seasonal reproductive cycle of the Texas tortoise is described in chapter 9. Clearly, there is an annual cycle of activity and inactivity (chapter 8), and this cycle may be largely controlled by temperature. However, we note that tortoises commonly become inactive in fall, when days become shorter and before the temperature is continuously below the range in which tortoises can be active.

External cues are necessary to ensure the synchrony of an individual's reproductive preparations with those of other members of its population. Zug (1993) pointed out that early experiments on the effects of photoperiod on reproduction emphasized birds, and it was assumed for many years that photoperiod was the principal timing cue in reptiles. However, Licht (1972) reported that temperature was the most important timing cue for reptiles and that photoperiod either had no effect or was a secondary influence. Moll (1979) cautioned that photoperiod is known to affect the nonreproductive physiology of turtles and merits careful consideration in studies of reproductive timing cues. To date, there have been no studies on the role of photoperiod in determining the reproductive rhythms of Texas tortoises. However, egg production occurs largely in the period between the vernal equinox (March 21) and summer solstice (June 22), which suggests that increasing day length might favor egg production and decreasing day length might bring about the cessation of egg production.

8

Behavior

The greatest danger to a speculative biologist is analogy. It is a pitfall to be avoided—the industry of the bee, the economics of the ant, the villainy of the snake, all in human terms have given us profound misconceptions of the animals.

—John Steinbeck, *The Log from the Sea of Cortez*

Many aspects of Texas tortoise behavior are described in this chapter, including interactions with other tortoises, such as combat, courtship, and intrasexual behavior; and interactions with the environment, such as burrowing, nest construction, use of a home range, homing, and dispersion.

Activity

A tortoise is considered active when it is walking or is out from under cover but is not basking. Auffenberg and Weaver (1969) reported that Texas tortoises had two activity periods, morning and afternoon. This bimodal distribution (fig. 8.1) of activity is generated by extreme midday temperatures and is muted in the early spring and late fall. Nonetheless, more tortoises (about 70%) are active in the afternoon. We attributed increased activity in the afternoon to the fact that temperatures at that time would be decreasing, while those in the morning would be increasing, frequently reaching the critical thermal limit tolerated physiologically (Rose and Judd 1975). The pattern, however, may hold even on cloudy days. Increased activity is noted after rains, or on mornings after rains, but if rain and cloudy conditions persist for several days, activity is reduced. Quite probably, increased activity after some environmental perturbation, such as rain, caused Grant (1936) to speculate about mass migrations. Tortoises have limited "reasons" for movements, such as searching for food or mates, and they

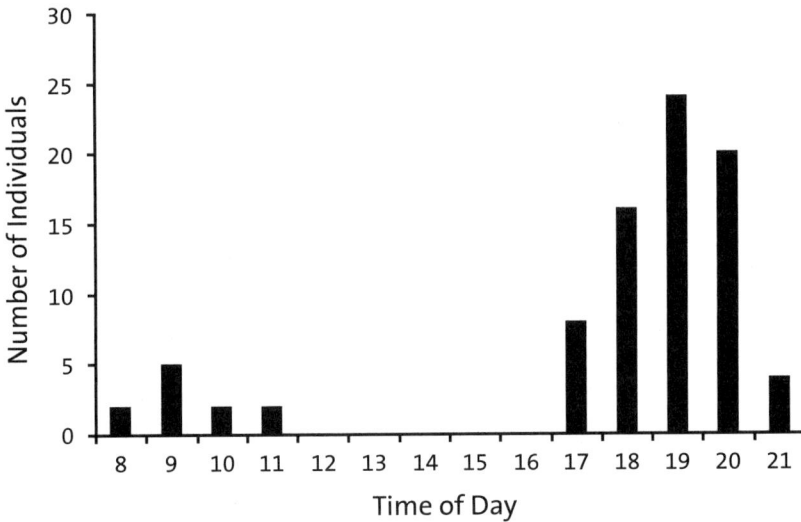

Figure 8.1. Bimodal activity pattern of the Texas tortoise during a period of high temperatures (July), with more individuals active in the late afternoon. The bimodal pattern is muted when midday temperatures are lower, as in early spring and late fall.

are not active after dark; thus, light does play a part in determining their activity periods in that it must be present. Lower temperatures probably trigger the increased activity that we see in the fall. If captive tortoises that are digging nest holes in which to lay their eggs are overtaken by darkness, they will cease nest construction and resume in the morning, sometimes returning to the same site.

Burrowing, Digging, and Nest Construction

Texas tortoises do not dig large earthen tunnels in which to retreat, but the other species of *Gopherus* dig extensive, energy-expensive tunnels that form a focal point for their activities. These burrows are as wide as the tortoise is long, so that it can turn around along its length. Texas tortoises will scratch out a shallow pan (see plate 6), usually at the base of vegetation or in a disruption in the soil surface such as a trough. As a general rule the pallet will be no deeper than about one-half the length of the tortoise, unless the pallet is exposed. On Loma Tio Alejos we found pallets that housed two and three tortoises, but we want to emphasize that such tunnels are exceptional.

Pallets are not an individual tortoise's resource and are occupied on an opportunistic basis. In friable sand habitats, little may be gained by devoting energy to pallet development because there are a significant number of holes and crevices in which to hide. Even so, on occasion, tortoises are observed in pallets in soft-soil areas.

Whereas pallets are sculpted with the forelimbs, nests are constructed with the hind limbs. How a site is selected is not known, but it is usually at the base of a plant, frequently under cactus. The female moves as though she is searching for an object, with her nose frequently touching the ground. She begins by rotating about five degrees several times, with her plastron resting on the ground and her forelimbs flicking surface soil backward. She usually releases urine and begins flicking the soil with half-circle motions of her hind feet. The neck of a constructed chamber is about one inch in diameter and about one inch deep. The neck opens into a larger chamber so that in cross section the nest appears like a vase whose cavity is off-center. It is remarkable how the female positions the eggs inside the chamber to maximize packing. Bear in mind that she cannot observe the process and that her legs are short, seemingly far shorter than is necessary to construct such a nest. In extremely hard, dry soils the female may wallow out a cone-shaped depression into which she lays her eggs (plate 27).

It takes about four hours for a female to dig a nest chamber, lay eggs, and cover the site. After each egg is passed, her head extends while she takes a deep breath; fluid seeps from her eyes. Once laying has ceased, she drags soil toward the hole with short sweeps of her hind limbs and finishes the process by flicking dirt with her forelimbs, tamping down soil with her elephantine hind feet to produce the cap, and urinating on the surface. This blending of the nest site with its surroundings is remarkable, for soon there are no visible signs of it. It might be that the female's tactile senses enable her to determine when the nest chamber is full. If she has dug a hole that will accept only two eggs, but she has produced three, then perhaps she dumps the other on the surface, and that accounts for the high numbers of uncovered eggs that are observed in this species. If you are long-lived, losing one-third of your reproductive effort for one year must be balanced against the energy that would be lost in constructing another hole, especially if urine is no longer available to soften the soil.

The role that urine plays in nest construction and protection needs further study. Undoubtedly, fluids soften the soil and thereby ease construction, for some substrates would be impenetrable by the tortoise without this aid. Some aquatic turtles store considerable water in their cloacae, which they use to soften soil during egg laying bouts (Cagle 1950; Mahmoud 1968). In dry years, if they have to construct more than one hole and in doing so deplete their water, they traverse the long distance back to water and try again another day. Texas tortoises use urine to soften soil and to flatten and compact the covering plug, but they do not have an unlimited supply of fluid. In addition, the protein components of urine probably serve to initially bind soil particles a little more tightly. Whether the fluid deters predators is questionable; Patterson's (1971b) experiments indicate otherwise. Because a pet dog did not eat meat soaked in three-day-old desert tortoise urine until five minutes had elapsed is not proof that urine serves an antipredator function. Tortoise urine that possesses a noxious odor or taste implies that the tortoise consumed some potent substance that passed through its kidneys, or was converted to such a component via the liver and excreted by the kidneys, or was converted in the bladder by some unknown reaction. Texas tortoise urine has a distinctive odor and color when tortoises are consuming prickly pear fruit, but the odor is not unpleasant, and we can safely vouch that it is bland to the taste, having had ample experience. Egg predators are motivated to pursue their high-energy seasonal payoff, and it is unlikely their well-honed senses could be overcome or thwarted unless grave unpleasantness was associated with the urine. For example, raccoons readily accessed nest chambers of aquatic turtles even though the nest chamber necks and caps had been treated with habanero and serrano chili peppers or 1,4-dichlorobenzene (mothballs) (Rose, pers. data). Habanero chilies have 350,000 to 600,000 Scoville units, enough capsaicin to cause simultaneous contractions of lung tissue and to collapse the urinary bladder of a human. Odor is probably the final sense triggered in a sweep-searching predator as it narrows the gap between itself and the resource.

As mentioned elsewhere, Texas tortoises do not dig nests at night. One assumes that there is a critical physiological point at which the neurological impulse to lay is not countered by the inducement of inactivity by darkness, and that females will continue digging until egg deposition takes place. However, the process of covering the eggs is long and arduous, and a tortoise already physically challenged from digging and laying might be more prone to predation while exposed.

Sand flicking is a general behavior observed in all members of the genus *Gopherus*. The instinct to dig is deeply ingrained, frequently resulting in sporadic backward sweeps of the forelimbs. This behavior probably accounts for the apparent paucity of sand at the mouths of gopher and Bolson tortoise burrows when compared to the volume of dirt that should have been removed from such a tunnel. Even small Texas tortoises scratch constantly if maintained in a container, and adults are also ceaseless in their scratching.

Combat

The degree of sexual dimorphism in the Texas tortoise raises interesting questions about sexual selection. Berry and Shine (1980) addressed this subject and concluded that most terrestrial turtles were combative, and thus, males were larger than females. Traditionally, sexual selection takes one (or both) of two routes: (1) male-to-male conflicts, and (2) female choice. The larger size and gular extensions of male Texas tortoises imply that combat is a significant component of their behavioral repertoire. Weaver (1970) categorized combat encounters, which occurred only between males, as either Type 1 or Type 2. In Type 1 combat, both males are combative. They approach each other from the front (plate 28), and after a few "pushes," one or both suddenly escalate the encounter on raised forelimbs and use one of their gular extensions in an attempt to push the other backward. The trick is for one tortoise to position a gular extension below that of the other tortoise, so that when the forelimbs are raised, the thrusting tortoise can bulldoze under the other. Conventional wisdom posits that the gular extension is used to flip or turn the other tortoise onto its carapace, but that view is too simplistic. The extension serves as a ram only in the sense that it allows one tortoise to come in under the other and use the full force of the anterior portion of its carapace to thrust under the adversary, pushing him up and backward, and causing him to roll onto his carapace. The intensity of these encounters varies with the season and the temperature, and they can be persistent if two tortoises are of equal size. However, a smaller tortoise that is sufficiently motivated with pumping hormones can be a formidable foe. On occasion, one or both tortoises bite the forelimbs, anterior carapace rim, or the gular of the other. These are significant bites and may cause the bitten tortoise to withdraw its head and forelimbs, making it more vulnerable to being pushed around and dominated. If one combatant is flipped, the dominant tortoise extends its head and forelimbs and frequently gives vigorous bites to the exposed limbs of the vanquished. This causes the "loser" to withdraw its head and cease attempting to right itself. Observers of the vanquished being bitten ask why the tortoises are so mean. They are not mean; they are simply—and beautifully—tortoises. We assure the reader that if the tables were turned, which they frequently are, the new victor would initiate the same punishment: only humans are mean. When one tortoise "cuts and runs," it appears quite dedicated to the project and may actually look comical as it bumps into objects and scampers away. The dominant tortoise may pursue the other while holding its head high, and its gait may be somewhat stilted if the intent is to not overtake the other. If the intent is to continue combat, then the loser will be pursued until it finds shelter or outpaces the oppressor.

On occasion there is a great deal of show when two equally large males face off and stand head-to-head with limbs, neck, and head maximally extended to make them as tall as possible. They may hold this pose for several minutes, with occasional weak thrusts. This behavior is apparently associated with establishing dominance or codominance based on size and may end in vigorous combat.

Inflicted bites are significant, but they do not generally penetrate the armor of large adult tortoises. On smaller tortoises, however, we have noted significant damage to the forelimbs (plate 29). Although two combatants may appear comical, like two small tanks doing battle on some dusty plain, we assure the reader that the participants are serious. On occasion the skin attached to the carapace above the neck becomes detached, providing an entry point for microorganisms and an egg-laying site for flies.

In Type 2 combat, only one of the pair is aggressive. The agonistic tortoise attacks the oppressed in a manner similar to how it would behave toward a female during courtship. The instigator bites more frequently than in Type 1 combat. The attacked tortoise attempts to withdraw its limbs while rotating on its axis, moving its head away from the action. It has the options of fleeing or burying its head and forebody in a pallet or under debris, but if it remains still for several minutes, the attack will stop and it will be allowed to go on its way without further harassment. The aggressive behavior occurs during the courtship period; however, Weaver (1970) did not observe males mounting dominated males.

Combat in the other species of *Gopherus* is also aggressive, but sexual dimorphism is weak. However, it is easy to see that male Texas tortoises are agonistic during the courtship period, implying that the more dominant males may well inseminate more females. However, because sexual size dimorphism is so weak in the other species, it appears that in female Texas tortoises mate choice is a driving force for this dimorphism. See more on this subject below.

Courtship, Mating, and Head Bobbing

In most any observation of Texas tortoises, and of other tortoises as well, the issue of head bobbing comes up. Almost whimsically, males initiate a series of head-bobbing motions when encountering another male or female. The bobbing intensifies during the mating season and has a deep genetic component. The intensity of the movements is affected by temperature; they go from slow, metronomic vertical motions at low temperatures to almost frantic bobs with lateral extensions at higher temperatures, especially in the presence of multiple females. In slow motion, the movements seem well orchestrated; however, they vary among individuals and, as mentioned, vary within the same individual with changes in temperature. Because intense head bobbing occurs during male-male as well as male-female interactions, it is logical to assume the bobs are signals of recognition (Auffenberg and Weaver 1969; Weaver 1970) and the intent to engage. The degree of variation in Texas tortoise head-bobbing sequences probably reflects the species having not been sympatric with other species of *Gopherus* for thousands of years, lessening the fine honing of the behavior as a species-specific indicator. Eglis (1962) summarized this tortoise behavior, but a detailed comparative study of the head-bobbing sequences of all species of *Gopherus* over a reasonable thermal range is warranted.

There is no way that a male Texas tortoise can inseminate a female without her full cooperation. The stage is set when the male attempts to subdue her by ramming and biting the rim of her shell; he has to be able to force her to remain quiescent and cooperative. She has to be successfully challenged and dominated, and the larger the male the more efficiently he is able to accomplish this task. He trails her and then rams her, usually from the rear or side, and then approaches her from the front, where he commences his biting routine, usually on the rim of the carapace or on the forelimbs. She turns away from him, and as he follows he tries to maintain a face-to-face position, but her arc is shorter than his and he has to scramble to keep up. Usually his chin is close to the ground and his neck is noticeably bent to the side as he twirls around. Eventually he will attempt to mount her from the rear, but if she is not subdued, all she has to do is walk away. If she is receptive, or at least interested, she twists on her axis, usually about 45°, or she may make a complete circle. He has to maintain his position and does so while drumming his hind feet as though he is making short vertical jumps. His neck is extended and his head is bent downward, with open mouth (plate 30). He may make audible hissing noises and frequently draw his head inward, where the subdentary glands strike the gular projections, causing the glands to release their products. In some older female Texas tortoises, the last vertebral scute (plate 31) may be visibly worn as a result of many instances of males rubbing and bumping their anal plates. Grant (1936) was the first to note this wear pattern in female desert tortoises. Because not all mature females exhibit this wear pattern, we conclude that some females are more acceptable to large males than others. If a female is receptive, she lowers the posterior portion of her plastron and opens her cloaca. At some point during this process, urine is released from one or both participants, wetting the ground in a circle and leaving traces of uric acid crystals. Without her cooperation at this stage, there is no way that the male can inseminate her, and all she has to do if the sequence is not to her liking is take a couple of steps forward and walk off. For some reason, when this occurs, although from a human perspective the male appears chapped as he is left standing on extended limbs with a whimsical look, he rarely pursues the female.

So, courtship in the Texas tortoise, like that in humans, is not a guarantee of copulation. To evaluate this assertion we counted how many times we observed courtship (mounted males) in free-ranging and captive tortoises and how many times we were able to observe intromission. There were 124 such courtship observations and 24 intromissions. We then measured the sizes of the males where intromission was observed. Only males with carapaces over 160 mm (6.3 in) long were observed in copulation (see fig. 5.5). These observations suggest that females will tolerate and participate in courtship with small males (especially if the females are small), but that on average it is the larger and more aggressive males that do the copulating. We thus posit that the selection force maintaining the sexual dimorphism in the Texas tortoise is female mate choice of the larger, more physically dominant males, who not only must establish themselves via male-male combat rituals, but must also be able to physically subdue females.

The interplay of several factors accounts for why the Texas tortoise exhibits striking sexual dimorphism and the other *Gopherus* species do not. It is the smallest of the five species, and it lays on average the fewest eggs (2–3 as opposed to an average of 6 for gopher and Bolson tortoises—but bear in mind that some gopher tortoises may lay as many as 25). Larger female gopher tortoises have larger clutches (Landers et al. 1980; L. Smith 1992), but we did not find this relationship for the Texas tortoise (Judd and Rose 1989) (see fig. 5.1). Female Texas tortoises are so small that they are constrained in reproductive output because of their internal volume. One way around this conundrum is to produce smaller eggs, but then the success of the hatchlings is compromised, especially in a harsh thorn-shrub environment. Calcium may be a limiting growth factor for females because they require so much of this element for egg-shell production. This is obvious from the thin-walled skeletal elements of a female carapace. It is easy to suggest that females, in an attempt to maximize the reproductive success of a low number of eggs, compromise by not developing thickened skeletal elements. Males, on the other hand, are not constrained by the loss of available calcium to egg construction. The longer gular extensions and thickening of the plastron in males can be interpreted as reinforcements for combat. Perhaps the small size of females demands a more exaggerated plastron concavity in males to facilitate coupling. The structural enhancements and anatomical modifications necessary to accommodate coupling with females, along with aggressive courtship and combat behaviors, may partially account for the interspecific differences in the degree of sexual dimorphism. This is certainly an area in need of further study.

Intrasexual Behavior

We have not observed male-male or female-female reproductive behavior in the field. We have observed such behavior several times in captive tortoises when the sexes are separated during the courtship period. In these encounters, the male acting out the female role and the female acting out the male role show all the requisite behaviors. When mounted, a male acting as a female twists on his central axis in one direction and then the other, not making any effort to escape the mounted male, which he could easily do by walking forward. A female assuming the role of a male is just as daunting as a male in her biting and chase behavior, and when she mounts, she rakes the carapace below her with her claws as she hops on her hind limbs, adroitly following the twisting motion of the underlying female. Only small males assume the role of a female, and only large females assume the role of a male. We have observed one female that exhibits male behavior routinely copulating with males and laying fertile eggs.

Reproductive behavioral sequences in males and females are dramatically different and are much more complex in males. It is intriguing how both behavioral sequences may be neurologically ingrained in individuals of each sex, and how the behaviors can be inappropriately triggered. We offer the following hypothesis: At low population densities and with sex ratios dramatically skewed toward females, a female assuming the role of a male might trigger ovulation. If sperm were stored from a previous copulation, then fertilization and zygote formation could ensue. It is conceivable that females assuming a male behavioral role also trigger ovulation in themselves, or else there would be no selective advantage to the pseudomale.

What value does the female accrue in acting like a male? Perhaps, as with whiptail lizards (Crews and Moore 1993), pseudocopulation is a reciprocal endeavor: you scratch my carapace and I will scratch yours. We see no value in males assuming the reproductive role of a female. Because combat and the initial stages of courtship are so similar, combat behavior exhibited by young males may be a mechanism of honing skills that will be used later in life. Like other males that initiate combat behavior toward inanimate objects, they are just confused: perhaps practice makes perfect!

The biting and ramming that males direct toward females certainly seems excessive. There is no doubt that a female Texas tortoise, if given the option, will not copulate with a male that is smaller than she, nor will she mate with one that cannot subdue her. She controls which male copulates with her. Once the ovum is ovulated and enters the ostium of the oviduct, it is quickly surrounded by dense albumen that sperm cannot penetrate. Slightly farther down the oviduct the albumen is coated with a fibrous matrix membrane, which is in turn enveloped with calcium. Fertilization, then, has to occur in the upper reaches of the oviduct prior to albumen accumulation. If the female ovulates early, before stored or ejaculated sperm are available, her clutch is lost. Evolutionarily, this waste of gametes is too large of a negative burden to tolerate, especially because there are so few eggs produced per clutch. We posit that the intensity of the male's courtship behavior, the level of his physical abuse, and his size are all ovulatory inducers. When ovulation occurs, stored sperm, if any, from the cloaca, as well as newly injected sperm, must be on their way to the point of fertilization. If the copulating male can stay connected for a sufficient time and inject sufficient viable sperm, he can be virtually assured that his half of the DNA component will be incorporated into the young. This race is not always won by the newest inseminator, however, as evidenced by multiple paternities in gopher tortoises (Moon et al. 2006).

Home Range and Movements

Animals move to obtain resources such as food, water, shelter, basking sites, or egg-laying sites. To study movement, one must be able to identify individuals. Usually this involves marking the animals. Filing notches or drilling small holes in the marginal scutes of tortoises might serve this purpose. To discern movement patterns or the size of the area used, marked individuals may be recaptured in subsequent surveys and capture locations plotted on a grid or map, or marked individuals may be tracked continuously using attached radio transmitters or devices that unwind thread as the tortoise moves. Today, Global Positioning System (GPS) tracking units are a tremendous aid to discerning movements and greatly enhance data analyses.

Individual animals may confine their movements to a given area where they obtain food, water, shelter, and mates. This area, if not defended from conspecific intruders, is termed the home range. Home ranges can overlap or they can be exclusive. A portion or all of the home range may be defended against conspecifics. This defended area is referred to as a territory. Home ranges often vary among seasons or years and are often correlated with animal size or age. They frequently differ between the sexes and may also vary with population density and food availability. There is a tendency for home ranges to increase in total area through time as different areas are used. The total area should then be treated as a life range.

In our experience, Texas tortoises apparently nest within their home range; however, males move out of normal activity areas to search for females, and there might be strong support for these movements to be included in the home range, but this is a subject of serious debate. We consider the departure from the home range for the purpose of reproduction to be a brief, periodic, oscillatory, directed movement that is predictable across years. This movement with its associated behaviors is utterly different in kind from the daily movements and patterns of behavior exhibited within the assumed home range. Without an understanding of sexually based behaviors, we will obtain spurious results from comparing the home range size of males and females.

Grids are established physical areas with marker posts at regular intervals in two dimensions, and they can be used to study a small animal's home range. The posts allow investigators to pinpoint the position of an organism within the confines of the grid boundary. Each row is lettered and each post within a row is numbered. This allows investigators to determine the space used, and, after some mental gymnastics, the space needs of the organism(s) being studied, which may be of primary importance when developing conservation paradigms. While mammalogists championed the grid concept early on, those studying lizards were not far behind. The reason is that small organisms use less space, and the confines of a reasonably sized grid can easily accommodate a significant number of home ranges and territories, allowing a researcher to determine if there are differences in how the sexes (or juveniles) use space through time. The data generated from the grid allow the researcher to test various methods of estimating population sizes, survivorship, and growth rates. It seems reasonable, then, that the size of the grid should be large enough to accommodate a statistically large enough number of organisms to mathematically facilitate the questions being asked, but not so large as to hamper thorough searches. With the availability of radio telemetry and GPS, grids are not as popular as they once were, but an established grid does allow an investigator to search an area more systematically than if he or she were randomly searching.

To address the effectiveness of line transects, we selected 10 search times and plotted the position of each tortoise on a model of the Yturria Ranch grid. We then determined 10 random lines across the grid for each search time and asked, how many of the known tortoise positions fall within 10 feet of either side of a line? Our results showed that of an average of 13 (8–19) known tortoise positions, only 6 (4–10) fell within 10 feet of either side of a line, providing a detection rate of 0.46. We would have greatly underestimated the population size of tortoises on the grid had we opted for this time-saving estimation method. One criticism of any study of an organism's use of space is that the investigator does not know where the organism is between sightings, and this holds for telemetry as well. In addition, space-use studies violate one of the basic tenets of a scientific study—there is no control, so the researcher does not know how capturing, marking, and placing a telemeter on or inside an organism modifies its behavior over time. We placed telemeters on 10 tortoises in Cameron County; all the animals dispersed from the capture site immediately, and only one was ever observed again.

All methods used to estimate home ranges have faults. Envision an organism walking about with a string-trailing device attached. The animal, if a tortoise, uses only a quite narrow area, say six inches on either side of the string. If an animal is marked, and captured several times, then the area encapsulated by capture points along a line that is drawn to the outside capture points allows one to estimate a home range (computer programs do this quite nicely). If points are connected to form the smallest area possible, the resulting plot is called a *minimum polygon*, and if the connected points maximize the size of the plot it is called a *maximum polygon*. If a mean capture point is calculated, the distance of each capture from the mean is ascertained, and then the mean of those data is calculated, and a circle can be drawn around the mean in which two standard deviations will encompass 98% of all points. This method, called the *probability density function*, has several obvious drawbacks, because it assumes a flat, circular home range. Today, various computer-assisted methods are employed, some rather sophisticated. The area used by our theoretical animal, however, is actually the length of the line times 12 inches, which will be considerably less than the estimated area. The same is true if one uses radiotelemetry. This technique allows one to find the animal, but the data are treated as before, because each time and place the tortoise is located represents a capture point. Telemetry, while considered a more sophisticated approach, has several problems; notably, there is no control and one does not know with assurance what organisms without telemeters would be doing. When evaluating home range data, one has to eliminate the possibility that the organism is moving randomly, that is, that the pattern is a random walk and there is no tendency for the organism to return to a given place.

To eliminate the possibility that Texas tortoises were moving randomly within our grid, we calculated home range sizes versus the number of times captured until there was no increase. Had the home range sizes increased with subsequent captures, it would have implied that the home range concept did not apply, with one caveat. Because the grid (and surrounding search areas) constituted a fixed space, calculated home range sizes would become asymptotic. To compensate for this problem, we generated a program that calculated the area around random points generated for 20 subjects captured 3 to 20 times (fig. 8.2). If Texas tortoises were nomadic and moving randomly, their calculated areas of

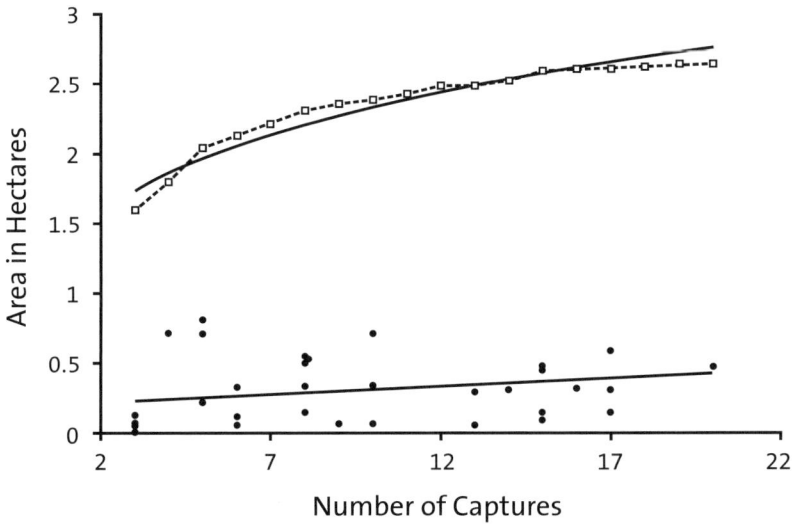

Figure 8.2. Models for determining if the general area of movements differs from what is expected if the movements are random. The upper line represents the mean from 3 to 20 computer-generated random captures, run 10 times each. For clarity, variances are not plotted. The upper line is asymptotic because the grid and boundary search areas are fixed, and the estimated home ranges cannot be greater than the area of the study. The variances follow the mean line and narrow with the number of generated captures. That the areas generated by the minimum polygon method (solid dots) fall well below the computer-generated line and do not increase with increased captures past 4 indicates that some Texas tortoises remain in the same areas for extended periods (years).

use would approximate the computer-generated line. In all cases in which Texas tortoises were captured four or more times over the years, their areas of use were well below the computer-generated line and its confidence limits. Four of the five species of *Gopherus* construct extensive burrows, and one would expect burrows to be activity centers. Auffenberg and Weaver (1969) may have reached their conclusion that Texas tortoises are nomadic because the tortoises do not construct burrows and appear to use pallets opportunistically.

One can also look at movements, that is, the distance between each capture, providing that capture times are relatively consistent. Those individuals that make extraordinary movements away from their activity areas are an inherent problem. When their mean movement data are calculated, some would say that these exceptional movements unduly influence the calculated mean. Perhaps the geometric mean, where the mean distance moved between captures is calculated as the square root of the product of the shortest and longest distances moved, would be more appropriate. We do not think that extreme exceptional

Table 8.1. Comparison of Texas tortoise mean home range size among males, females, and juveniles, calculated using four different methods

Method	Mean home range size (ha)		
	Males	Females	Juveniles
Minimum polygon (8 or more captures/individual)	0.45	0.22	...
Maximum distance between capture points	0.78	0.46	0.54
Mean distance between captures	1.01	0.55	0.54
Density probability function	2.38	1.40	2.35

Source: Data from Rose and Judd (1975).

movements should be minimized, because they are a biological component of the organism in question, and extreme minimal movements are seldom eliminated in the calculations; however, extreme movements are certainly outside the home range concept. Then there are those who discount the whole concept of home range and insist that if one backs out far enough into space to observe movement, all mobile organisms will be perceived as occurring in a limited area of activity.

We used four methods to estimate home range size (density probability function, minimum polygon, mean distance between successive recaptures, and maximum distance between capture points) over a two-year period (Rose and Judd 1975). All four methods of estimating home range size indicated that males have larger home ranges than females (table 8.1). These findings contrast sharply with those of Auffenberg and Weaver (1969), who indicated that no correlation existed between sex and distance traveled and that the tortoises were perceived as being nomadic, maintaining restricted activity ranges for only a few days. Our analyses showed that some Texas tortoises had home ranges for up to two years. Indeed, of those tortoises captured two or more times, 95% of the males and 100% of the females were captured less than 60 m from their activity centers. In addition, we found a strong negative correlation between female size and home range size (r = -0.7036, $p < 0.01$), but no significant correlation between male size and home range size (r = -0.1552) (Rose and Judd 1975). Individual variability was the single most important factor in determining home range size.

In another study, we reported on the home range of the same population after five years of observation (Judd and Rose 1983). We used two methods of estimating home range size, the density probability function and the minimum polygon. Tortoises captured four or more times were used for home range estimations. There was no significant difference in the home range size estimated after two years or five years of study in either sex using either of the two methods of estimation (table 8.2).

A major difference from our 1975 study was that there was no correlation

Plate 18. Posterior view of bones of a female Texas tortoise showing holes in the eighth costals where the pubic bones have worn through. Image by Don Anders.

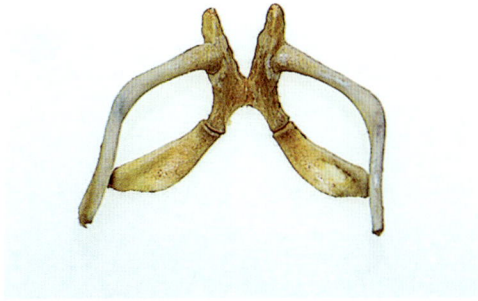

Plate 19. The hyoid apparatus of a Texas tortoise. The four cornua and the basihyoid serve as a basket to anchor the tongue and the muscles involved with swallowing and expanding the anterior aspect of the throat.

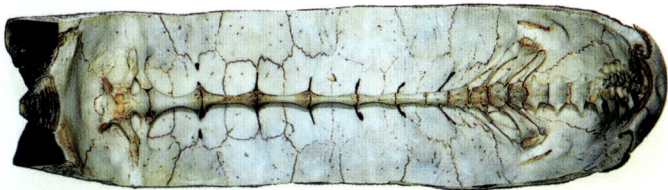

Plate 20. Cutaway view of the carapace emphasizing the neural elements and the small rib remnants (black) in the thoracic region.

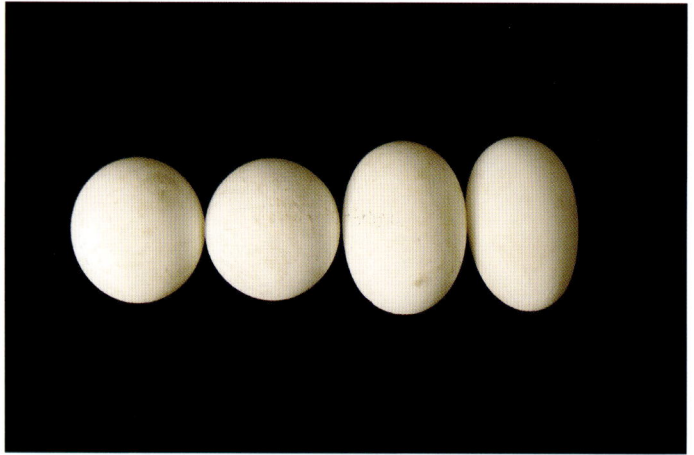

Plate 21. Variation in the shape of Texas tortoise eggs. Image by Don Anders.

Plate 22. A newly laid Texas tortoise egg with the upper half of the shell removed to show the rich yellow-orange yolk and the thick albumen.

Plate 23. Hatchling Texas tortoise showing the remaining yolk sac and the transabdominal groove where the tortoise was folded while developing in the egg. The fold generally flattens in one to two days.

Plate 24. General color of a juvenile Texas tortoise (named Root-hog), showing the yellow scute centers and edges of the marginals. Note that the fourth and fifth toes on the forefeet were worn off as a result of the hatchling digging through compact, dry soil to escape the nest bowl. But alas, a raccoon killed him.

Plate 25. Large male Texas tortoise from Cameron County, Texas, showing the uniform horn color of older males.

Plate 26. Group of captive Texas tortoises at feeding time. All of the tortoises routinely retired inside the safety box at night and during temperature extremes. The opening was sealed during the winter.

Plate 27. Female Texas tortoise laying eggs in a conical chamber. Typically tortoises lay eggs in a nest chamber connected to the surface by a short neck. The cone-shaped cavities are the result of efforts in extremely hard-packed soils; the females are adept at constructing classic cavities in such soils if sufficient urine is available.

Plate 28. Two large male Texas tortoises attempting to intimidate each other. The individual on the right shows dark phase coloration as well as the mild templing of the scutes typical of captive animals. He was 10 years old at the time the image was made, a product of the Silver Spoon Effect (see chapter 5), and he is also the hatchling in plate 23. The other male may be his parent.

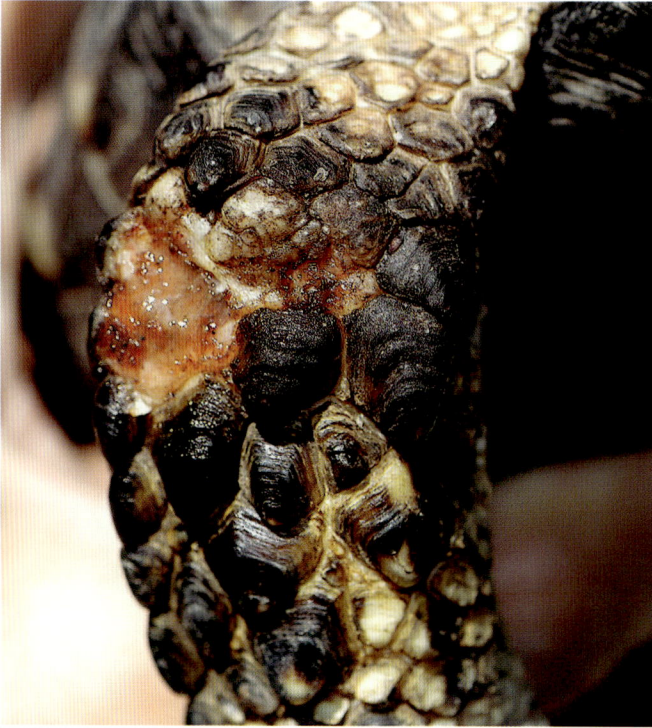

Plate 29. Bite damage on the forelimb of a male Texas tortoise. While such injuries are rare, once the integument of the forelimb is breeched, continuing combat may result in more serious injuries, like those induced by predators.

Plate 30. Male Texas tortoise (T-bone) mounted on a female. Note the male's open mouth, the urine-soaked soil, and the white uric acid crystals. Both participants may express urine, but females may be the primary source. Because the urine of the Texas tortoise is somewhat slimy, it may serve as a lubricant.

Plate 31. Worn twelfth marginal scute of a female Texas tortoise. The wear is attributable to friction generated by males rubbing against her with their xiphiplastrons as they drum with their hind feet.

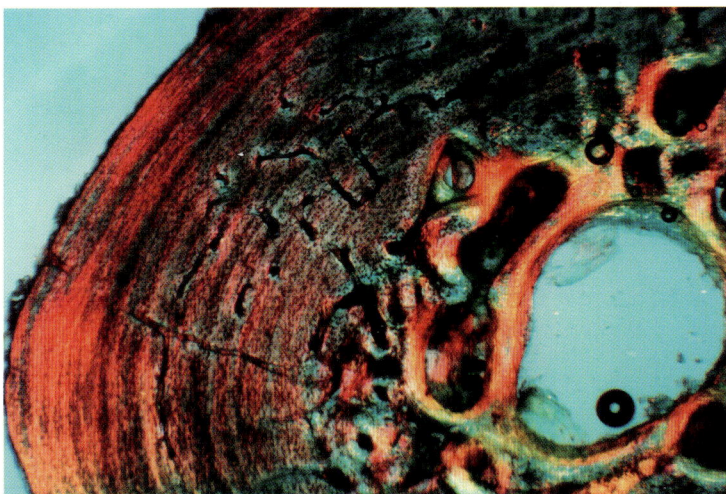

Plate 32. Stained bone section of a femur from a Texas tortoise. Because of wear, it is basically impossible to ascertain the ages of Texas tortoises by counting scute rings, except in young animals. Bone sections of the skeletons of individuals of known size may offer an alternative method, assuming that the number of rings represents age. Image by David H. Johnston.

Plate 33. Texas tortoise habitat destruction in South Texas.

Plate 34. Invasive buffelgrass has become prevalent over much of the range of the Texas tortoise. During droughts the grass poses a serious fire danger for all manner of organisms. In addition, it limits the movement and potential food sources of Texas tortoises.

Table 8.2. Comparison of Texas tortoise home range size using two different methods and data collected over 2- and 5-year periods

Sex & time period	Minimum polygon			Density probability function			
	N	Mean (ha)	SD	N	Mean RR (m)	SD	Mean size (ha)
Males - 2 yr	16	0.29	0.25	19	41.0	22.0	2.38
Males - 5 yr	20	0.47	0.38	20	47.2	21.6	2.57
Females - 2 yr	13	0.17	0.19	17	26.3	19.8	1.40
Females - 5 yr	15	0.34	0.27	15	29.9	18.6	1.42

Source: 2-yr data from Rose and Judd (1975); 5-yr data from Judd and Rose (1983).

Note: RR = recapture radius.

between tortoise size and home range size in either sex after five years of study. The method used to estimate home range size makes a large difference in the size of the home range (table 8.2). This is because the density probability function method assumes a circular shape for the home range and the minimum polygon method does not. The home range size determined by the minimum polygon method is strongly correlated with the number of captures up to some point, while the density probability function method is not influenced by the number of captures.

Kazmaier et al. (2002) reported home range estimates for 36 tortoises from the Chaparral Wildlife Management Area in Dimmit and La Salle Counties, Texas. Thirteen females and seven males were on sites grazed by cattle, and nine females and seven males were from ungrazed sites. Kazmaier et al. calculated home range size using five methods and reported that home ranges were larger for males than for females using all methods of calculation. They also concluded, using the 100% minimum convex polygon method, that tortoises in ungrazed pastures had larger home ranges than tortoises in grazed pastures. We present their data for the 100% minimum convex polygon method in table 8.3.

We find that the data exhibit no significant difference between the sexes within a treatment (grazed pastures: $t = 1.741$, 18 df, $p > 0.1$; ungrazed pastures: $t = 1.702$, 14 df, $p > 0.1$). If data for the two treatments are combined, the sample size is increased and males have significantly larger home ranges than females ($t = 2.644$, 34 df, $p < 0.02$). The difference between sexes appears to be largely attributable to the very large home range estimates for some males in the ungrazed pastures. Table 8.3 shows that the home range for the seven males extends from 9.2 to 130.7 ha (22.7 to 323 ac). Thus, the largest home range is 14.2 times the size of the smallest. Unless one knows the dates that tortoises were found at the locations used to calculate the home range, it is impossible

Table 8.3. Texas tortoise home range sizes using the minimum convex polygon method

Site Status	Gender	Sample size	Range (ha)	Mean	SE
Grazed	Female	13	1.0–19.8	5.0	1.4
Ungrazed	Female	9	1.5–21.6	6.8	2.1
Grazed	Male	7	4.8–23.2	9.5	2.4
Ungrazed	Male	7	9.2–130.7	31.8	16.6

Source: Data from Kazmaier et al. (2002) in the Chaparral Wildlife Management Area in Dimmit and La Salle Counties, Texas.

to assess whether these locations likely represent a home range or the dispersal and relocation of an individual, but the latter interpretation might be correct.

T-tests of the data in table 8.3 also show no significant difference in home range size between treatments for either sex (males: $t = 1.330$, 12 df, $p > 0.2$; females: $t = 1.036$, 20 df, $p > 0.2$). If the sexes are combined, the sample size is increased, but the home range size is still not significantly different between treatments ($t = 1.657$, 34 df, $p > 0.1$).

The home range sizes reported by Kazmaier et al. (2002) for tortoises from Dimmit and La Salle Counties are markedly larger than those we reported for tortoises from Cameron County (Rose and Judd 1975; Judd and Rose 1983). Population density is greater in Cameron County and tortoises are larger there as well. The smaller home range size, greater population density, and larger body size of the Cameron County tortoises may be a consequence of a more abundant food supply and higher-quality food in Cameron County compared with Dimmit and La Salle Counties. Food abundance and quality has not been measured at the two locations, but visual inspection suggests greater vegetation cover in Cameron County as well as greater rainfall and relative humidity. Kazmaier et al. (2002) suggested that home ranges in Cameron County might have been small because searching occurred on a 3.3 ha (8.2 ac) grid. However, we regularly searched a boundary area of 36.5 m (120 ft) around the grid to account for immigration and emigration, and we periodically searched beyond this boundary area as well (Rose and Judd 1975; Judd and Rose 1983).

Tortoises are patchily distributed in Cameron County and are largely limited to lomas where upland vegetation occurs. Surrounding habitat is salt marsh or salt flats and inappropriate for tortoises. Indeed, cordgrass (*Spartina*) marshes make tortoise locomotion very difficult. Confinement of tortoises to lomas may contribute to higher densities and smaller home range sizes there, but food abundance and quality must be great enough to support the higher densities. Available habitat in Dimmit and La Salle Counties is more continuous and may contribute to larger home range sizes. Obviously, this aspect of Texas tortoise biology is a fertile area for further research.

We reported that capture points for some tortoises approximated a circle (Judd and Rose 1983). To test whether tortoises used the area within a circle containing all capture points equally, we divided this circle into a central core and three outer rings, with each of the four divisions having equal area. Only tortoises captured four or more times were used in the analysis. Capture points for tortoises were clustered within the central core of the estimated home range. In fact, 54.4% of male captures and 64.1% of female captures were contained within the central core. No other ring had more than 18.3% of the capture points for either sex. Clearly, the tortoises do not use all of their home range equally. Rather, most of the activity is confined to a central core area. Male tortoises do not patrol and defend a perimeter; thus, no part of the home range is a territory. This observation is corroborated by the fact that pallets are used on an opportunistic basis (Rose and Judd 1982).

Temporal and spatial changes in home range might occur with the gradual expansion of the area used or with relocation of the area of activity. The gradual expansion of area is not supported for the Texas tortoise because there were no significant differences in mean home range size based on two years of data and five years of data (Judd and Rose 1983). Males may shift their areas of activity, because 37% of the tortoises that were marked only once were males.

Homing

Homing is the ability of a displaced individual to return to an original location. When a turtle moves out of its home range to lay eggs and then returns, the implication is that it has a way of finding its home again. Homing ability in turtles has been most studied in sea turtles. Relatively little is known about homing ability in terrestrial turtles. However, Metcalf and Metcalf (1970) reported that ornate box turtles (*Terrapene ornata*) returned home from a displacement of up to 2 mi (3.2 km). They suggested three mechanisms by which ornate box turtles find their way: (1) visual recognition of landmarks, (2) olfaction, and (3) celestial navigation. Auffenberg and Weaver (1969) suggested that Texas tortoises use visual and olfactory cues to orient themselves when moving within their activity ranges. Chelazzi and Delfino (1986) showed experimentally that washing the nasal epithelium of Hermann tortoises (*Testudo hermanni*) with a 2.0% solution of zinc sulfate impaired the ability of tortoises that were displaced 500 to 1000 m (1640 to 3281 ft) to orient toward home and to return to the vicinity of their capture locations. After 14 days, only 1 of 26 (3.8%) treated tortoises returned home, while 9 of 24 (37.5%) controls did so. Homing ability has not been tested in the Texas tortoise or in any other species in the genus *Gopherus*. However, Hellgren et al. (2000) reported that female Texas tortoises carrying radio transmitters commonly made extensive forays out of their resident home range during the nesting season. This suggests that females have homing ability. Studies are needed to test homing ability in males, females, and juveniles and to discover the sensory modalities used if the tortoises are able to home. On one occasion in October 1993, tortoise tracks associated with a single linear movement, observed in a sandy roadbed, were followed for over 180 m (590 ft) in the Chaparral Wildlife Management Area.

Hamilton (1944) reported a mass migration of Texas tortoises in which he observed 16 individuals over two to three miles of highway in South Texas in 1938, and a like number just off the highway. By today's standards, that would be a large number of tortoises, but Hamilton's observations were made on August 3 after a significant rain, a time of maximum courtship activity. Therefore it is less likely that all movements were in a single direction.

Dispersion and Dispersal

Dispersion refers to the distribution of individuals in space. Three patterns of dispersion are possible: (1) individuals may occur in groups (clumped), (2) they may be randomly distributed, and (3) they may occur with uniform spacing among individuals. We divided the Yturria Ranch grid into 25 quadrants (36.5 m by 36.5 m [120 ft by 120 ft]) and plotted the location of activity centers (determined using the density probability function method) (Rose and Judd 1975). We then tested the dispersion of activity centers for randomness using the Poisson distribution and found that the activity centers were highly clumped ($t = 140.394$, 24 df, $p < 0.01$). The distribution of tortoise activity centers was closely associated with the distribution of prickly pear, that is, there was a high frequency of tortoise activity centers in quadrants having relatively large clumps of prickly pear. Indeed, 62% of the tortoises observed on the grid were under or in the immediate vicinity of prickly pear. Thus, the clumped distribution of the tortoises may reflect the clumped distribution of prickly pear and the dependence of the tortoises on it for food, shade, and protection. Clumped dispersion may also result because during the reproductive season males track females and stay close to a given female for days at a time. The males engage in courtship and copulation attempts at frequent intervals during the days they track a given female. Consequently, it is relatively common to find a male and female together during the reproductive season.

Dispersal is often used to refer to any sort of one-way movement between locations, but some suggest that it is probably a one-way movement to a location unknown to the dispersing individual. Often dispersal occurs in juveniles as they move away from the home range of the parent. We reported that none of 13 juveniles marked on our Yturria Ranch grid in 1972 were present in 1976 (Judd and Rose 1983). We suggested that juvenile Texas tortoises in Cameron County disperse from their hatching sites and do not establish home ranges until they become sexually mature. Kazmaier et al. (2002) interpreted large movements by males between recaptures from juvenile to adult classes as evidence that Texas tortoises in Dimmit and La Salle Counties exhibit male-biased juvenile dispersal. The longest movement was 11.3 km (6.9 mi) for a three- to four-year-old juvenile over an elapsed time of 368 days, but a five-year-old male moved 8.0 km (4.9 mi) in 22 days. They reported that movement directions within each of the five sex/age categories they examined were random.

Dispersal may serve to reduce competition and decrease the probability of inbreeding. Costs of dispersing include the potential of being caught out in the open with no shade as temperature increases to a lethal point, increased exposure to predation in unfamiliar areas, and the risk of being unable to find adequate food and other resources in an unfamiliar area.

Habitat Selection

Kazmaier et al. (2001a) examined habitat selection of tortoises at the Chaparral Wildlife Management Area in Dimmit and La Salle Counties using nine habitat types: (1) blackbrush, (2) escarpment, (3) hogplum, (4) paloverde, (5) parkland, (6) woodland, (7) riparian, (8) white brush, and (9) old-field. They concluded that Texas tortoises selected habitats with a broad range of woody canopies while avoiding riparian and old-field habitats. Thus, tortoises tolerated a broad spectrum of habitat types. Parkland was more highly ranked in ungrazed sites, and woodland was more highly ranked in grazed pastures.

9

Population Ecology

Human science fragments everything in order to understand it, kills
everything in order to examine it.

—Leo Tolstoy, *War and Peace*

This chapter examines some of the dynamics of Texas tortoise population ecology, including sex ratios, age structure, mating season, clutch size and frequency, egg retention, reproductive potential, and the male reproductive cycle.

Sex Ratios

We reported that the sex ratio of a population of Texas tortoises from eastern Cameron County was essentially 1:1 (Judd and Rose 1983). We marked and released 102 tortoises (43 males, 30 females, 3 adults of unrecorded sex, and 26 juveniles) on a study grid and its boundary area over a five-year period from 1972 through 1976. The greatest difference in any given year between the number of males and females was in 1973, when there were 24 (61%) males and 15 females on the study area. Sex ratios among juveniles were unknown because there is no way to reliably establish the sex of live immature individuals.

The number of females from year to year was remarkably constant, that is, in 1975 there were 16 females on the study area and in all other years there were 15 females. One might think that this constancy in number of females was because the same females were captured, but this was not the case. Comparison of the coefficient of similarity between 1972 and 1976 showed that only 57.9% of the females were the same. The similarity of males between 1972 and 1976 was 45.8%.

Auffenberg and Weaver (1969) showed that in two of the three populations they studied, the percentage of adult females was double that of males. Most turtle species have a 1:1 sex ratio (Auffenberg and Iverson 1979), but asymmetric ratios are known in other tortoises, for example, desert tortoises (Lucken-bach 1982). The island-like nature of lomas in eastern Cameron County may result in reduced gene flow between the lomas, even though they are in close proximity, and contribute to variation in sex ratios and other demographic parameters (Auffenberg and Weaver 1969; Rose and Judd 1982). It is difficult to ascertain in a long-lived species with an asymmetric sex ratio what factors might have moved the ratio away from 1:1 without knowing when the perturbations occurred or whether they were gradual. If Texas tortoise sex ratios are 1:1, then our suggestion may be incorrect that females are at a selective disadvantage compared to males because of the physiological rigors of egg production and nest construction with its concomitant increased exposure to predation, or, if this suggestion is correct, there may be an unequal sex ratio of hatchlings that favors females. This aspect warrants renewed and vigorous investigation.

Hellgren et al. (2000) reported that the overall sex ratio of a population in Dimmit and La Salle Counties did not differ from 1:1, with there being 261 males and 243 females. The sex ratio changed, however, from female dominated at younger adult ages (4–6 years old) to male dominated at older ages (7–15 years old). The inclusion of males without secondary sexual characteristics in the female count might temper the assumption of female-dominated samples at younger ages, because females were recognized by an absence of male secondary sexual characters. However, this possibility does not influence the male bias found in years 7 through 15. If we assume that once females begin to reproduce, they are at increased risk of predation and physiological stress associated with egg production and calcium depletion, then older males should dominate the sex ratio, as observed by Hellgren et al. (2000). However, we were not able to confirm that females in the Cameron County population (Yturria Ranch) had decreased survivorship (Rose et al. 2011).

Hellgren et al. (2000) concluded that male bias in older age classes existed because of a lower survival of females. They attributed higher female mortality to the smaller size of females at sexual maturity, rendering them more vulnerable to predation and the physiological costs of reproduction. They suggested that females were calcium deficient because they needed calcium for eggshell production, their diet was low in calcium, and their consumption of *Opuntia* fruit and cladophylls that are high in calcium oxalates reduced calcium availability. Calcium deficiency over time might lead to reabsorption of bone in the carapace until the pelvic girdle wears a hole in the overlying bone (see plate 18). Twisting of the pelvis during nest construction might contribute to wearing the bone away. High female mortality might result from this bone loss and the associated impaired movement and increased susceptibility to microorganisms, or it might result directly from calcium deficiency (Hellgren et al. 2000). However, while we have noted many females with worn carapace bone at the attachments of the ilia, we have never observed a tortoise in which the bone hollow was not covered with a scute. In addition, we have never found shells of dead tortoises in which the opening was not covered, nor have we found a scute with holes associated with a dead tortoise. The carapace is no doubt thin, but we posit that scute production probably outpaces bone degradation. In addition, some male carapaces exhibit these openings.

Age Structure

We found that juveniles made up 25.4% of a Cameron County population (Judd and Rose 1983). Hellgren et al. (2000) reported that 34% of the population in Dimmit and La Salle Counties were juveniles, and Kazmaier et al. (2001b), working with the same population, found that juveniles made up 32.1% of the population in grazed pastures and 37.2% of the population in ungrazed pastures. Obviously, some skeletochronological work is needed to establish age structure (plate 32).

Mating Season

Mating and activity seasons for males are essentially synonymous. We reported that tortoises are inactive in December, January, and February (Rose and Judd 1975). Males may court females from March into November, but courtship is observed most frequently from May through October. However, intense courtship and copulation generally occur from July through August. Females exhibit little overt mating behavior such as head bobbing, but they may signal receptivity by olfactory means. Auffenberg (1969) reported that the cloacal secretions of *Gopherus* females enable males to identify their gender, and it might be that these secretions also convey information on the females' receptivity. In addition, individuals may wipe secretions from their subdentary glands onto their forelimbs to indicate receptivity to courtship (Auffenberg 1969). We conclude that the sensory cues used in courtship and recognition are in need of more study.

Clutch Size and Clutches per Year

Clutch size does not have a universally accepted definition, and this lack of agreement causes confusion. Dunham et al. (1988) distinguished between primary, secondary, and tertiary clutch sizes. Primary clutch size is the number of enlarged, yolked, preovarian follicles that are likely to be ovulated in a single nesting bout; secondary clutch size is the number of enlarged follicles that are actually ovulated; and tertiary clutch size is the number of fertile eggs in the nest. Thus, different estimates of clutch size may be obtained by counts of enlarged follicles, corpora lutea, oviductal eggs, or eggs within the nest (Dunham et al. 1988). Determination of clutch size can be further complicated because turtles may retain their eggs internally over long periods of time (Cagle and

Tihen 1948) and divide their complement of eggs between more than one nest (Fitch and Plummer 1975). For example, painted turtles (*Chrysemys picta*) in Michigan retain eggs for two and a half to three weeks (Gibbons 1968), and stinkpots (*Sternotherus odoratus*) in Wisconsin may retain eggs for five to eight weeks (Edgren 1960). We suggest that clutch size is best regarded as the number of eggs yolked and shelled at one time rather than the number of eggs deposited in a nest.

Clutch size data for Texas tortoises vary among studies. Grant (1960) reported a clutch size of 6 eggs for a captive female. Mean clutch sizes calculated from data in Auffenberg and Weaver (1969) and Rose and Judd (1982) are 1.4 and 4.3 eggs, respectively. Hellgren et al. (2000) used ultrasound to determine clutch size of an inland population in Dimmit and La Salle Counties and found mean clutch size to be 2.07 (range 1–4) for 49 females.

We showed a significant positive correlation between carapace length and clutch size that suggested that older (and larger) females produce larger clutches than younger (and smaller) females (Rose and Judd 1982). We also designed a study (Judd and Rose 1989) to ascertain the egg production of Texas tortoises, to resolve the marked differences in estimates of clutch size, and to discover the clutch frequency in South Texas by x-raying marked females in the field at regular intervals throughout two reproductive seasons.

Adult females were radiographed in the field at approximately two-week intervals from May 20 to July 29, 1986, and from May 18 to August 7, 1987, to determine clutch size, egg size, and clutch frequency. Females were radiographed once a month from August to October 1986, March to April 1987, and September to October 1987. We found that clutch size ranged from 1 to 5 in both 1986 and 1987. Only 2 of 29 females bore 5 eggs. Mean clutch size was virtually identical in the two years. In 1986, clutch size averaged 2.6, and in 1987 mean clutch size was 2.7. There was no significant difference in the means ($t = 0.215$, 27 df, $p > 0.5$). Five females produced clutches in both years of the study (table 9.1). There was no significant difference between years in clutch size ($t = 0.250$, 8 df, $p > 0.25$), although none of the females had the same clutch size in the two years.

To examine the relationship between clutch size and date during the laying season, we used only the first radiograph and date that a female exhibited maximum clutch size; that is, we did not use subsequent radiographs for the same female when clutch size was the same or when the female had already laid a portion of her clutch. The regression equation was not significant ($t = 0.676$, 27 df, $p > 0.5$). Thus, clutch size did not change as the reproductive season progressed.

Variation in clutch size estimates of Texas tortoises are shown in table 9.2. Grant's (1960) data can be interpreted as indicating a clutch size of six eggs or clutch sizes ranging from one to six eggs. Auffenberg and Weaver's (1969) data are based on "eggs per nest" rather than on the examination of clutch size. Their estimate was lower than our estimate using radiographs (Judd and Rose 1989) and that of Hellgren et al. (2000) using ultrasound and was likely a consequence

Table 9.1. Comparison of clutch size among female Texas tortoises producing clutches in both 1986 and 1987

Female	Carapace length (mm)	Clutch size	
		1986	1987
1	188	5	3
2	166	3	4
3	161	2	3
4	175	1	2
5	182	4	2

Source: Data from Judd and Rose (1989).

Table 9.2. Variation in clutch size estimates of the Texas tortoise

Range	Mean	Authority
(6)	Single observation	Grant (1960)
1–3	1.4	Auffenberg and Weaver (1969)
3–7	4.3	Rose and Judd (1982)
1–5	2.6	Judd and Rose (1989)
1–4	2.1	Hellgren et al. (2000)

of females partitioning clutches in time and space by laying in more than one nest. Our higher mean estimate (Rose and Judd 1982) is based on necropsy of females killed on highways by cars. The difference between this estimate and our subsequent estimate (Judd and Rose 1989) may be a chance event due to small sample size, an excellent year for food resources, or difficulty in reconstructing the exact number of shelled eggs present because they were crushed when the tortoises were killed. As a consequence, we now believe that six to seven eggs in a single clutch would be excessive: there is just not enough space inside the female tortoise.

Clutch Frequency and Egg Retention

Prior to our study (Judd and Rose 1989), data on clutch frequency (Auffenberg and Weaver 1969; Rose and Judd 1982) were based on inferences from the laying dates of captive and wild females and necropsy of road-killed females. There were no data from sequential captures of marked females. Consequently, existing clutch frequency data were open to more than one interpretation. For example, Auffenberg and Weaver (1969) suggested that there were two peaks of egg laying—one in late June and July and another in late August and September. They stated that some females might lay four clutches per year. We found that 3 of 11 females with shelled eggs in April also had large ovarian eggs (Rose and Judd 1982), and we suggested that if shelled eggs were laid in May or June it was feasible for large ovarian eggs to be ready for laying by late summer.

These data do not provide compelling evidence that females lay multiple clutches per year. The fact that some captive females laid eggs twice in a summer does not necessarily mean that the eggs were from two different clutches: they could have been portions of a single clutch laid at different times. Indeed, there is evidence to support this alternative. The greatest number of eggs that we found in a nest was four (Rose and Judd 1982), but the largest clutch size of a necropsied female was seven (but note that we are no longer comfortable recognizing a clutch size of seven eggs). This suggests that females with large clutches laid eggs in more than one nest. Grant (1960) provided further evidence: he reported that a captive female laid three eggs in a nest on June 8 and later deposited single eggs on the surface on June 19, July 9, and July 13. Because only 11 days elapsed between June 8 and 19, it is unlikely that the single egg laid on June 19 was from a second clutch of eggs. Likewise, because only 4 days passed between the laying of single eggs on July 9 and 13, it is improbable that they were from different clutches.

We sought definitive information on clutch frequency by x-raying marked females at two-week intervals throughout the reproductive seasons of 1986 and 1987 (Judd and Rose 1989). Of the 77 females we radiographed, 29 had eggs (see fig. 1.5). Three radiographs were required of a female to confirm the production of two clutches. In 1986, 10 females were radiographed three or more times and in 1987, 14 females were radiographed three or more times. None of the 24 females was found with shelled eggs, then without eggs, and then with eggs again. There was evidence that 16 females produced only a single clutch of eggs. Sixteen females (10 in 1986 and 6 in 1987) were radiographed with shelled eggs and then later without eggs. Apparently all 29 females found with eggs produced a single clutch. This was to be expected because of the energy, water, and calcium requirements for a small tortoise to produce multiple clutches of such relatively large eggs.

The percentage of females not producing eggs was estimated from females radiographed two or more times in a given year at least one month apart. Six of 16 females (37.5%) in 1986 and 5 of 14 (35.7%) in 1987 did not have shelled eggs. These data imply that not all adults produce eggs each year. This is an important observation because it impacts the calculation of fecundity and thus the development of management paradigms.

Seven females radiographed three or more times provided data on egg retention. Two females retained at least a portion of their shelled eggs for 39 days, and one female each retained for 37, 36, 35, 23, and 21 days. Two of these females provided evidence of clutch partitioning, that is, they showed a reduction of egg number in successive radiographs. A female with five shelled eggs had one egg 39 days later, and a female with three eggs had one egg 13 days later.

Auffenberg and Weaver (1969) concluded that mature females might lay up to four clutches a year. Based on the simultaneous presence of shelled eggs and large ovarian eggs in females killed on highways in April, we suggested that some females might lay two clutches per year (Rose and Judd 1982). Turner et al. (1986) provided conclusive evidence for one to three clutches per year in desert tortoises in California. We emphasize, however, that three clutches per season by desert tortoises are rare; the overwhelming generalization is one or two clutches per season (Turner, pers. comm.). Morafka (1982) reported that captive Bolson tortoises laid multiple clutches.

We found no evidence of multiple clutches in Texas tortoises from South Texas (Judd and Rose 1989). It is possible that the seven females that we concluded were retaining eggs for considerable periods actually laid a clutch and then developed exactly the same number and sizes of eggs in a second clutch. However, given the variation that existed within individuals in clutch size in successive years (table 9.1), it seems unlikely. Another possibility is that the two females that we interpreted as laying a portion of their clutches might have laid all of the eggs in their first clutch and then produced a second clutch of one egg. Turner et al. (1986) reported that the interval between the first oviposition and the appearance of second clutches in desert tortoises ranged from 9 to 19 days. Because we did not attempt to x-ray a given female at an interval shorter than 14 days, it is possible that we missed the times when these two females had no eggs (Judd and Rose 1989).

Because we did not find evidence for two clutches—that is, we did not find a female with eggs, then without eggs, and then with eggs again—we proposed that the simplest interpretation is that females retain shelled eggs for considerable periods of time and partition the clutch in time and space (Judd and Rose 1989). A comparison of clutch size and the number of eggs in a nest provides evidence for this interpretation. The largest number of eggs we found in a natural nest was four, but we also reported clutch sizes of five, six, and seven eggs (Rose and Judd 1982; Judd and Rose 1989). Further evidence is provided by the sequence of laying dates that Grant (1960) reported for a captive female, that is, three eggs were laid in a nest on June 8, and single eggs were laid on the surface on June 19, July 9, and July 13. One might conclude that all six eggs were from the same clutch and were partitioned in time. Alternatively, if one uses Turner's (1986) data on time for clutch development for desert tortoises, two other interpretations are possible. There could have been a clutch of three eggs laid on June 8, followed by the development of a second clutch of three eggs that was laid one at a time on the dates given above. Finally, there could have been an initial clutch of three eggs, followed by a second clutch of one egg and a third clutch of two eggs that were laid singly four days apart. Regardless of which interpretation is selected, partitioning of the clutch in time occurred. Ever mindful of Occam's razor, we realize that the simplest interpretation is that there was a single clutch.

The number of clutches per year varies among several species of aquatic turtles (Gibbons 1982; Gibbons et al. 1982; and others). Thus, annual differences in clutch frequency of Texas tortoises may be revealed in the future. The shelled eggs that Auffenberg and Weaver (1969) found in females in September certainly suggest the possibility of two clutches in some females in some years.

Single annual clutches appear to be the usual case for gopher tortoises. Iverson (1980) reported that females from northern Florida laid one clutch per year. Likewise, Landers et al. (1980) reported that gopher tortoises from Georgia nested no more than once annually and that a given female was successful in producing young an average of only once in about 10 years.

Hellgren et al. (2000) calculated that the Texas tortoise population in Dimmit and La Salle Counties had a clutch frequency of 1.3 clutches per female. They used the length of the period when gravid females were found and divided that by 30 days, which they assumed would be the time between clutches, to arrive at the estimate of clutch frequency. Thus, while they have no direct evidence that females lay more than one clutch of eggs, their estimates suggest that some females in this inland population lay two clutches per year.

Reproductive Potential

Assuming a population of 100 females in which 36.6% do not produce eggs in any given year, and assuming that 2.6 eggs are produced per clutch and that there is one clutch annually, then a total of 168 eggs would be produced in one year in the Cameron County population (Judd and Rose 1989). If 60% of the eggs hatch, as reported by Judd and McQueen (1980), 101 hatchlings would result. This estimate is no doubt much higher than reality because it does not account for egg predation. There are no data on survivorship of hatchlings to additional years in the Cameron County population.

The mean percentage of females identified as gravid by ultrasound in the Dimmit and La Salle Counties population was only 34.3% (Hellgren et al. 2000). This, coupled with a mean clutch size of 2.0 eggs per female and a clutch frequency of 1.34, would result in 100 females producing 95 eggs. If only 60% of the eggs hatch, 57 hatchlings would be produced in a year, about half the productivity of the Cameron County population.

Male Reproductive Cycle

Nothing is known about the spermatic cycle of Texas tortoises. Behavioral observations indicate that copulation is most frequent from late July into September, so active sperm production is probably initiated in June. Cloacal lavages of captive males in June revealed active sperm; however, a few active sperm were observed as early as April. To establish at what age and at what point in the season spermatic activity is activated requires access to tissue from animals of known age, tissue that cannot be obtained without killing tortoises. At this stage, the ends do not justify the means! Captive males at two years of age have been observed actively courting their female siblings.

10

Environmental Resistance
to Tortoise Survival

Nature's polluted.
There's man in every secret corner of her
Doing damned, wicked deeds.
—Thomas Beddoes, *Death's Jest-Book,*
or the Fool's Tragedy, act 2, scene 3

In this chapter we see that a newly hatched Texas tortoise is not necessarily guaranteed to live to a ripe old age. There can be nest failure, juvenile mortality, limited survival to sexual maturity, and other challenges to long life expectancy.

Nest Failure and Failure to Hatch

Judd and McQueen (1980) reported hatching success for Texas tortoises of 60% and 61.5% in successive years. Auffenberg and Weaver (1969) suggested that predation is heavy on Texas tortoise eggs, as indicated by their finding of 88 disrupted nests, but they mentioned that some of those might represent successful hatches. In our experience, eggshells from successful hatches rarely reach the surface. Hellgren et al. (2000) reported they were unable to locate a single nest after four full seasons of radio transmitter field monitoring of 15 to 20 females.

Judd and McQueen (1980) suggested that predation might be heavy on Texas tortoise eggs because the last egg in a clutch is often laid on the surface near the nest, allowing predators to easily locate the nest. The fact that females often lay in more than one nest and on different days suggests that predation on eggs might indeed be high, and that clutch partitioning in space and time might reduce such predation. However, the deposition of eggs on the surface above the nest does not seem prudent, unless the odor of the underground eggs

141

is masked while a mesomammalian predator consumes the yolky surface egg on-site. Also, it could be that once a predator eats an isolated egg, its olfactory acuity is lessened for a short period. However, this assumption blithely suggests that predators are not well honed at their sustenance tasks, an assumption one retains at one's peril.

We reported finding a large number of partially covered eggs and eggs deposited on the surface with no apparent attempt at nest construction (Judd and Rose 1989). Generally, when a partially covered egg was found, two eggs were found in the shallow chamber below, suggesting that the nest cavity could not accommodate three eggs. Single eggs laid on the surface might represent remnants of a clutch that was deposited after the chamber was covered. In 1986 and 1987, we found four eggs together on the surface two times, three eggs once, two eggs seven times, and single eggs on nine occasions (Judd and Rose 1989). No excavation attempts by predators were noted.

Females are known to use urine to soften the ground during nest construction. Perhaps when water is limited, nest construction in hard soils is affected. However, we reported that although there was considerable rain in 1987 in the area of our grids, 26 eggs were found on the surface that year (Judd and Rose 1989).

Landers et al. (1980) reported that gopher tortoises frequently constructed nests in the mound of soft dirt at burrow entrances. Turner et al. (1986) reported desert tortoise females at Goffs, California, laying eggs at the bottom of a burrow. Over much of their range, Texas tortoises have to dig egg chambers most often in hard, compacted soils. Strecker (1928) reported Texas tortoise eggs laid on the surface of a "flinty outcrop" that was too hard for nest construction, but this incident bears scrutiny. Hard-packed soils would also provide a problem for hatching tortoises. Brode (1959) reported that only half of the hatching gopher tortoises in his study in Mississippi dug their way out of their nests. He attributed this to the hard-packed overburden. In 2011, we observed a hatchling Texas tortoise that had worn off the lateral digits of its forelimbs while exiting a nest constructed in dry, hard-packed soil.

Both the Yturria and Reed Ranch grids were on hard-packed soils (Judd and Rose 1989). Comparative studies in more inland areas (with softer soils) might help elucidate the degree to which substrate affects oviposition and the loss of fully formed eggs. We predicted that in soft soils, female Texas tortoises would construct more nest chambers and lay one to three eggs per chamber. In hard soils, the energy demands of constructing a single chamber might be not only great but life threatening to a female whose bladder water was depleted. If a female cannot wallow a conical chamber in the soil with her posterior carapace as she digs her nest, her legs are not long enough to construct a chamber that will hold more than two eggs if the soil is resistant. Perhaps this partially explains why "extra" eggs are left on the surface.

Juvenile Mortality

We did not find any dead juveniles in our Cameron County population, but none of the 13 juveniles we marked on the Yturria Ranch grid in 1972 were present in 1976 (Judd and Rose 1983). This might reflect dispersal rather than mortality. Likewise, Hellgren et al. (2000) did not capture many juveniles and consequently did not present survivorship data for tortoises younger than five years. They noted that life history parameters of adult females in their population necessitated a 24.5% survival rate of hatchling tortoises to four years of age in order to maintain a stationary population. Using the same life history information for females, they estimated survival from egg laying to one year of age at 52.8% and stated that only 5 out of 1,000 hatchlings would survive to age 20.

Auffenberg and Weaver (1969) reported that skunk and raccoon tracks in soil near destroyed nests implicated these two mesomammalian predators as tortoise nest predators. They suggested that southern plains wood rats (*Neotoma micropus*) might predate eggs and hatchlings. They identified foxes, coyotes, skunks, bobcats, raccoons, and indigo snakes as probable predators of juvenile Texas tortoises. Dramatic increases in feral hog populations in South Texas in recent years warrant heightened concern about these apocalyptic omnivores because they have the capacity to consume even large tortoises.

Survival to Sexual Maturity

Hellgren et al. (2000) reported that the modal age at which female Texas tortoises attained sexual maturity in the population in Dimmit and La Salle Counties was five years. They showed life table data for three different clutch frequencies, 1.1/year, 1.3/year, and 1.9/year. Estimates of survival to five years of age for these clutch frequencies were 24.6%, 20.2%, and 14.2%, respectively. These data all assume a stable population where r = 0.0.

Annual Survival of Adults

Kazmaier et al. (2001b) reported annual survival rates determined from frequency-age distributions of tortoises estimated at 5 to 12 years of age based on counts of growth annuli of scutes. Annual survival estimates (sexes combined) were 82% for grazed pastures and 79% for ungrazed pastures. They also reported on adult survival based on radio tracking of 28 females and 19 males for average periods of between 534 and 668 days. Annual survival rates in grazed pastures were 84% for females and 73% for males. Annual survival rates in ungrazed pastures were 70% for females and 83% for males. Thus, two methods of determining survival rates yielded similar estimates.

Annual survivorship estimates for the Yturria and Reed Ranches in Cameron County, Texas, are shown in table 5.2 (Rose et al. 2011). Survivorship on

Table 10.1. Comparison of mean carapace lengths of female Texas tortoises from Cameron County and Dimmit and La Salle Counties, Texas

Location	Sample type	Authority	N	Mean (mm)	SE
Cameron County, Yturria and Reed Ranches	females with eggs	Judd and Rose (1989)	29	170.0	3.2
Cameron County, Yturria Ranch	adult size and larger	Rose and Judd (1982)	32	149.8	2.5
Dimmit/La Salle Counties	females 5 yrs and older	Hellgren et al. (2000)	221	140.9	1.5

the Reed Ranch was significantly lower than that on the Yturria Ranch. These study areas are only 6.4 km (3.9 mi) apart, but they differ in vegetation. Prickly pear is much more abundant on the Yturria Ranch (see section on regulation of numbers in this chapter). Known mortality supports the survivorship estimates. On the Yturria Ranch, 225 tortoises were marked, and 9 marked animals were found dead. Known mortality was at least 3.5%. On the Reed Ranch, 151 tortoises were marked, and 15 marked animals were found dead, yielding a mortality of 9.9%, which is 2.8 times greater than that on the Yturria Ranch.

Life Expectancy and Replacement

Hellgren et al. (2000) calculated that about 5 out 1,000 hatchlings reach 20 years of age in the population in Dimmit and La Salle Counties. If one uses growth rings to estimate age, a large proportion of adults in the Cameron County population appear to live longer than 20 years. However, after about 18 years of age (Auffenberg and Weaver 1969) or 24 years of age (Judd and McQueen 1982), growth rings are so close together that it is impossible to make accurate counts. Furthermore, growth rings in many old individuals are worn smooth and are impossible to discern. Nevertheless, it is probable that a much higher proportion of individuals is older than 20 years in the Cameron County population than in the Dimmit and La Salle population, because the former individuals have a higher number of discernible growth rings and longer carapaces than females of the latter population (table 10.1). Females with eggs in the Cameron County population were significantly larger than both adult-size females without eggs in that population, as reported by Rose and Judd (1982) ($t = 5.010$, 59 df, $p < 0.001$), and females five years old or older from the Dimmit and La Salle population, as reported by Hellgren et al. (2000) ($t = 6.767$, 248 df, $p < 0.001$). Additionally, females from Cameron County had significantly longer carapaces than females from Dimmit and La Salle Counties ($t = 2.191$, 251 df, $p < 0.05$), as reported by Hellgren et al. (2000). While we have inadequate data on growth rings of females from Cameron County, a large proportion had more than 18 growth rings. Certainly, Cameron County females were larger, which suggests that they were older than those from Dimmit and La Salle Counties.

Regulation of Numbers

The smaller size and earlier sexual maturity of Texas tortoises as compared to other extant species of the genus *Gopherus* suggest a life history strategy in which environmental variation affects adult survival more strongly than it does juvenile survival (Hellgren et al. 2000). Species following such a strategy would be expected to have relatively shorter life spans and to expend greater reproductive effort than congeneric species that do not follow this paradigm. A shorter life span may hold for the Dimmit and La Salle population, but it is not clear that it does for the Cameron County population. In addition, greater reproductive effort does not appear to hold for either population compared with other species of *Gopherus*. Small size carries with it the paradox of fewer eggs. Decreasing egg size and multiple clutches per year are ways around this problem, but Texas tortoises seem to not avail themselves of these two options. Females making extensive movements outside their home ranges to nest (Hellgren et al. 2000) could reduce intraspecific competition, implying that competition might be a factor limiting population density.

Auffenberg and Weaver (1969) found marked differences in the mean carapace length of tortoises on closely situated but isolated lomas. They concluded that these size differences were due to differential growth rates and suggested that variation in mean size among populations was probably due to differences in food and its availability. Because grass is present everywhere tortoises occur, but prickly pear abundance varies markedly, we reasoned that if the distance between food plants is great, tortoise movements should be greater, and tortoises should have larger home ranges where distances between prickly pear clumps are great. We also hypothesized that if prickly pear is a major food source, tortoise density should be greater in areas of abundant prickly pear. Thus, we established three hypotheses:

1. Tortoise size should be larger where prickly pear is more abundant.
2. Tortoise density should be greater where prickly pear is more abundant.
3. Tortoise home range size should be smaller where prickly pear is more abundant.

Because prickly pear was relatively abundant on the Yturria Ranch, we established another study area on the Reed Ranch, where prickly pear was scarce (table 10.2), in order to test these hypotheses. The Yturria and Reed Ranches are only 6.4 km (3.9 mi) apart, so they are exposed to the same climate and weather. Prickly pear is far more abundant in both absolute and relative cover on the Yturria Ranch (table 10.2).

Table 10.3 shows a comparison of tortoise density on the Yturria and Reed

Table 10.2. Comparison of total cover and cover provided by prickly pear on the Yturria and Reed Ranches, Cameron County, Texas

Category	Yturria Ranch	Reed Ranch
Total cover (trees, shrubs, cacti; %)	20.92	25.72
Prickly pear (% total cover)	6.12	0.44
Prickly pear (% relative cover)	29.27	1.70

Source: Data from Judd and Rose (1989).

Table 10.3. Comparison of Texas tortoise density on the Yturria and Reed Ranches, Cameron County, Texas

Parameter	Yturria Ranch	Reed Ranch
Grid area (ha)	3.3	2.0
Total tortoises marked	81	70
Maximum density (per ha)	24.5	35.0
Minimum density (per ha)	12.9	18.8
Mean Jolly estimate (per ha)	16.5	29.5

Source: Data from Judd and Rose (1983).

Ranches. Maximum density estimates assume that each of the tortoises marked on a grid was a resident. Minimum density estimates represent the mean number of tortoises per year known to be alive. Both estimates show density to be higher on the Reed Ranch. The Jolly stochastic method also shows density to be higher on the Reed Ranch. Thus, the hypothesis that density is greater where prickly pear is more abundant is not supported. The abundance of prickly pear did not affect the density of Texas tortoises.

Table 10.4 shows a comparison of home range size within sexes between the Yturria and Reed Ranch grids. We used only tortoises captured four or more times in calculating home range size. Home range size was smaller on the Reed Ranch in both sexes and significantly smaller for Reed Ranch females. Mean home range size of Reed Ranch males was less than half that of Yturria Ranch males, a relationship that was close to being significant (table 10.4). Clearly, the hypothesis that home range size would be smaller where prickly pear was more abundant was not supported.

There was no significant difference in the carapace length of males on the Reed and Yturria Ranches, but females on the Yturria Ranch were significantly larger than females on the Reed Ranch ($t = 2.848$, 52 df, $p < 0.01$) (fig. 10.1). On the Yturria Ranch grid, only 10% of females had carapaces shorter than 150 mm (5.9 in), while on the Reed Ranch, 50% of the females had carapaces shorter than 150 mm. For a food-related growth response to be expressed in females

Table 10.4. Comparison of minimum polygon home ranges within sexes between the Yturria and Reed Ranches, Cameron County, Texas

| Sex and location | Home range size (ha) | | | t | p |
	N	Mean	SD		
Males, Yturria	20	0.47	0.38		
				1.965	> 0.05
Males, Reed	11	0.23	0.18		
Females, Yturria	15	0.34	0.27		
				2.076	< 0.05
Females, Reed	8	0.13	0.12		

Source: Data from Rose and Judd (1975) and Judd and Rose (1983).

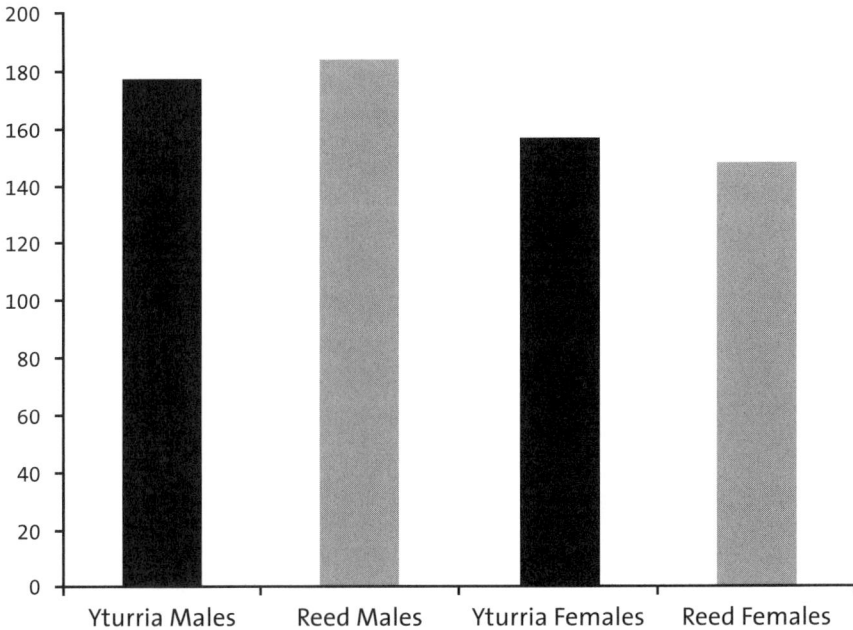

Figure 10.1. Mean carapace lengths of a sample of male and female Texas tortoises in two study grids, one with abundant prickly pear (Yturria Ranch; 49 males, 30 females), and one with scarce prickly pear (Reed Ranch; 30 males, 24 females). Although the difference between the females from the two ranches appears small, it is statistically significant.

but not in males would require that females depend more on prickly pear as a food source than males. We have no data to support or refute this statement. An alternative explanation is that the size relationships observed on the two grids reflect differential mortality among cohorts of females. Again, we have no data to support or refute this possibility. Regardless of the explanation, the hypothesis that tortoise size should be greater where prickly pear is more abundant is not supported for males but is supported for females.

Density estimates for the Dimmit and La Salle population range from 0.26 to 0.28 adults per ha (0.69/ac) (Hellgren et al. 2000; Kazmaier et al. 2001c), which are 46 to 72 times lower than our density estimates for the coastal Cameron County populations. Much of the difference in density estimates is likely real and reflects differences in the productivity of the two locations. However, some of the difference in density estimates is a product of methodology. For example, juveniles were included in the density estimates of Cameron County populations but not in the density estimates of the Dimmit and La Salle population. Also, Cameron County estimates were based only on habitat known to support tortoises, but the Dimmit and La Salle estimate apparently included habitats such as old-field and riparian areas that were avoided by tortoises (Kazmaier et al. 2001a).

11

Conservation, Monitoring, and Maintenance

We are like tenant farmers chopping down the fence around our house for fuel when we should be using Nature's inexhaustible sources of energy— sun, wind and tide. . . . I'd put my money on the sun and solar energy. What a source of power! I hope we don't have to wait until oil and coal run out before we tackle that.
—Thomas Edison, in a conversation with Henry Ford and Harvey Firestone, cited in James Newton, *Uncommon Friends: Life with Thomas Edison, Henry Ford, Harvey Firestone, Alexis Carrel, and Charles Lindbergh*

This chapter provides a look at many aspects of the Texas tortoise's relationship with humans, including the ways humans affect tortoise habitat and the many issues that arise when humans decide to maintain tortoises in captivity.

Historical and Legal Status

Tortoises are long-lived and produce few eggs per year. Their paradigm is based on individual longevity and putting out a high-quality, relatively large reproductive product that best fits their environment. To maintain population densities, exclusive of disease, nest predation, injury, general predation, and environmental perturbations, a single female has to produce but two new adults in her lifetime. Obviously, if significantly large numbers of breeding females were removed from the population, overall egg production would be lowered and population levels would decrease. For long-lived species at low to moderate densities, that decline might be slight annually but have lasting negative impacts. Assuming that a female tortoise lives 30 years and is producing eggs in her fifth year, and that she produces an average of 2.5 eggs per year, she will produce 63 eggs in her lifetime, and 97% of her reproductive effort will go into producing two adults.

Early on, pet suppliers removed significant numbers of Texas tortoises from the environment, and this, along with the animal's low reproductive output, prompted the Texas legislature to establish a law in 1967 protecting the tortoises from being injured, killed, collected, or possessed for "sale or barter or commercial exploitation" (Rose and Judd 1982). The United States now endorses the Convention on International Trade in Endangered Species of Wild Fauna and Flora (CITES). Foreign government permits are required under Article IV of the Convention before importing endangered taxa. All species of *Gopherus* were placed in appendix II of the List of Endangered Species, which protects not only endangered and threatened species but also those species that might be negatively impacted if strict regulations were not in place (Morafka 1982). The Texas legislature in 1977 developed nongame regulations that prohibit persons from taking, possessing, transporting, exploiting, selling, offering for sale, or shipping 81 species of Texas vertebrates including Texas tortoises. These regulations also apply to goods made from protected animals. The Texas Parks and Wildlife Department now lists Texas tortoises as "threatened." The U.S. Fish and Wildlife Service provides no such protection, probably because the natural range of the Texas tortoise in the United States is restricted to Texas, and state regulations are thought adequate for its protection.

Predation by Humans

There is little evidence that Amerindians used Texas tortoises for food, as humans currently use the gopher tortoise and Bolson tortoise. One might agree that protein is protein, but there appears to be no evidence that the Coahuiltecans inhabiting the inland areas of Texas consumed tortoises, although Newcomb (1980) commented that "there seems to be little that Coahuiltecans failed to eat which could be used by the human digestive apparatus." Their diet included what is known in anthropological terms as the second harvest, that is, seeds, primarily those of prickly pear, harvested from human fecal material and animal scat. Newcomb (1980) listed "land tortoises" as food items of the Tonkawas, but the known range of these people was north of what is considered the range of Texas tortoises. Perhaps the land tortoises Newcomb mentioned were the smaller, ornate box turtles (*Terrapene ornata*) (Judd and Rose 2000). One might ruminate on the reasons Texas tortoise shells are not found in Coahuiltecan middens (thin shells easily weathered after cooking or degraded as a result of rodents), but the possibility exists that they were not routinely eaten. However, Berlandier noted that they were gathered and eaten by members of his traveling parties in Texas.

While we have not verified it, we understand that illegal immigrants passing through the scrub country in South Texas eat Texas tortoises routinely. The tortoises are small and easily cooked over undetectable fires. Although the meal would be minimal, the liver would provide some nourishment, and when compared to no food, a small tortoise might take on the glow of a gastrointestinal banquet. Yet the idea of a tortoise tossed alive on its carapace onto the glowing embers to die and stew in its own juices curdles the neurons and provides intrusive images fit only for nightmares.

Texas tortoises are prepared for the Mexican tourist market by being dried and coated with a heavy layer of varnish. It is sad to see the grotesque image of a once-vibrant tortoise mounted vertically with a miniature bass fiddle attached as though it were playing in the dance band on the *Titanic*. As souvenirs, Texas tortoises become silent sentinels, celebrating the prurient whims of large-brained apes with small minds.

One robust population of Texas tortoises (one tortoise found every six minutes with four people searching) that we observed on several occasions was suddenly decimated, and extensive searches revealed no tortoises. Because turtle shells remain intact for considerable time after death (Dodd 1995) and we saw no evidence of a major die-off, we assumed that the tortoises' close proximity to a primary highway resulted in them being captured and whisked away to join many of their comrades in the pet trade or to be prepared as tourist gewgaws.

Undoubtedly, today, vehicles deliver the biggest negative impact on Texas tortoise populations. Aside from crossing roadways, the tortoises routinely graze along the right-of-way, where grass is thicker and shorter and there is less hindrance to walking. Unfortunately, this also places them in plain view to passing motorists, who run them over or pick them up and haul them off to an unregistered future.

Habitat Alteration

In the Lower Rio Grande Valley, less than 5% of native Texas tortoise communities remain (Jarhrsdoerfer and Leslie 1988). This destruction (plate 33) extends to northeastern Tamaulipas, Mexico, and Zapata and Laredo, Texas, where land is modified for agriculture or to improve grazing. Early in the nineteenth century, large-scale brush removal was begun in South Texas (Mutz et al. 1978; Inglis et al. 1986). As with many other endeavors, efficiency improved with time, and in the 1910s and 1920s trees were killed by girdling and with chemicals such as kerosene or diesel fuel. Tractors and Caterpillars wreaked havoc starting in the 1930s by pulling huge chains or cables between them, a process called *chaining*. Anything in front of a chain is destroyed. Plants are gouged from the earth, mauled, and tumbled. Then comes a mechanized rake that gouges six to eight inches into the sward and stacks debris in piles or windrows where it is left to rot or burn. A root plow—a massive blade—is dragged along after chaining and severs anything at a depth of six to eight inches, about where you would expect a tortoise to be in a pallet. The crown of horrors is a roller-chopper, which is much like a steamroller with massive offset blades. After such an insult, the cadaverous land is converted to improved pasture or is used for row crops. This process kills many tortoises; their bones provide ample evidence. Many are crushed and others are buried alive. Those surviving environmental scarring are left in a moonscape habitat that is devoid of cover and dissected with deep furrows and obstacles to movement. If the land is used for row crops, little can be reclaimed by tortoises; however, if pastureland is the goal and prickly pear is allowed to return, surviving tortoises may find themselves in suitable, if not ideal, habitat.

Today, chaining and root plowing continue at a rapid pace, but the introduction of herbicides in the 1970s (Inglis et al. 1986; Beasom and Scifres 1977; Mutz et al. 1978) gave rise to another dimension of habitat alteration, although the land is not disrupted as with chaining. We have no information on the short-term or long-term effects of these powerful chemicals on exposed animals, but as with chaining, herbicides cause shade cover to be severely reduced.

Brush control of semidesert scrub might not provide lasting agricultural benefits. Upsurges by grasses are probably due to the release of nutrients from disrupted soil, decaying stems, and roots (Jarhrsdoerfer and Leslie 1988). Chaining has to be repeated every 2 years, and root plowing every 15 years. Not surprisingly, mesquite density in treated areas was found to be three to four times greater 25 years after treatment and root plowing than it was in untreated areas (Fulbright and Beasom 1987).

There is no doubt that grasses can be so thick (plate 34) as to thwart tortoise movement, and this might account for why some tortoises are observed along roadsides. Bury and Smith (1986) suggested that controlled burns of grass might open up areas for tortoise activity and limit their need to use roadways. They suggested small controlled burns of less than 1 ha (2.5 ac) because of the known negative effects of burns on tortoises (Cheylan 1984; Stubbs et al. 1985). Introduced buffelgrass is a bane for tortoises because it has some nutrient content for cattle, which means that ranchers continue to sow its seeds. It grows in tough, tightly packed hillocks that effectively serve as one continuous, matted wall.

South of San Antonio, an estimated four million acres are surrounded by eight-foot "deer proof" fences. These fences are barriers to medium and larger tortoises, but smaller ones become embedded in the wire, where they languish and die or become easy targets for predators (Engeman et al. 2004). High mortality seems to abate with the age of the fence, such that after several months deaths are minimal. Fences, however, might impact local body size distributions, depending on mesh size.

Many tortoises are killed by vehicular traffic, and it seems reasonable to assume that this will not abate but will steadily increase. The Texas Department of Transportation (TxDOT) lists 32,798 miles of roadway in the counties in which

Texas tortoises are known to occur (the calculation includes a percent reduction in peripheral counties where tortoise occupancy is not expected throughout the county). This figure does not seem high initially until one realizes that it is approximately 10 times the east-west distance across the United States, or over 4 times the earth's diameter. In addition, in those very same counties the number of vehicles times the miles driven per day is 61,851,260. We point out that these are minimum figures because there are roadways not under TxDOT control, plus numerous ranch roads called "two-tracks" that penetrate deep into the heart of tortoise habitat. Many dead tortoises are found associated with these tracks, which become overgrown with grasses and block the view of a vehicle driver. This section would not be complete if we failed to mention mentally challenged sphenocephalic drivers who intentionally run over tortoises and other animals and have been observed to go out of their lanes to do so. One might wish those drivers to be reincarnated as a flea on a Hell Hound's back and be chased by a pair of red-hot fangs for 10,000 years.

Security fencing currently being constructed along the U.S.-Mexico border will destroy valuable habitat and will stand mostly as another monument to human stupidity. Currently the fence will capture some of The Texas Nature Conservancy's Southmost Preserve, which is located in a bend of the Rio Grande near the southern tip of Texas. This jewel of the Lower Rio Grande Valley encompasses one of the last sabal palm tree stands in the country and is part of the Lower Rio Grande Valley Wildlife Corridor. On the surface, it appears that tortoise populations inhabiting this area should be well protected; however, some of the area lies between the border fence and the Rio Grande and some of it is scheduled to be degraded. Destruction of any of this habitat and the specific destruction of sabal palms should open up several slots in the second ring of hell for those whose shortsightedness is responsible for this mass transgression against nature and the organisms tied to this space. So, if we assume that matters within the Lower Rio Grande Valley cannot get any worse for its unheralded inhabitants, we will be wrong.

Nothing has been published relative to how railroad tracks impact tortoise movements or mortality. There are not a lot of tracks south of San Antonio but their estimated combined length is considerable (1405 km [873 mi]). In addition, the tracks from San Antonio to Eagle Pass, from San Antonio to Laredo, and from Laredo to Hebbronville, Alice, and Corpus Christi pass directly through prime habitat. These could be formidable barriers but probably do not affect mortality.

The Eagle Ford Shale is a gas- and oil-producing formation of Cretaceous origin. It extends as a 50-mile-wide and 400-mile-long crescent from Mexico up into East Texas. Unfortunately, a great deal of the South Texas swath angles through prime Texas tortoise territory. Between the discovery of this formation in 2008, when 26 well permits were approved, and February 7, 2012, there was a 13,360% increase in approved permits (data from the Texas Railroad Commission). It has been estimated that by 2020, 5,000 wells will have been drilled. Increased drilling activity (276 active rigs as of March 9, 2012) requires more roads—more roads and traffic, more dead tortoises. This Grand Guignol has just begun, and we hope that those who sit up and take notice do not keep sitting; however, hope is not a plan and rarely trumps experience.

Subsidized Predation

General predators of Texas tortoises were covered in chapter 3. Plains wood rats are probably the primary egg predator (Auffenberg and Weaver 1969). These wood rats and the Texas tortoise are ecologically and mutualistically linked and occur in close proximity. Several factors, however, have favored increases in mesomammalian predators. Urbanization brings with it cats and dogs, many of which are fed outside and are allowed to roam freely. In addition, many people actually feed predators because they represent the reasons they moved away from the huddled masses and beehive existence in the city. Skunks, raccoons, foxes, coyotes, bobcats, and opossums readily capitalize on increased food resources. Some of these predators are considered cute, until they start consuming pets. Steep decreases in the prices of animal pelts in the 1980s released hunting and trapping pressures and resulted in vaulting predator population increases, with devastating results. Aside from food, subsidized predators take advantage of shelter, more numerous and safe nesting sites, and ease of dispersal.

The role of imported red fire ants (*Solenopsis invicta*) as a nest predator of Texas tortoises is unknown. Hard-shelled eggs deny entry (Diffie et al. 2010), but the long delay in the pipping process leaves an extended open window through which ants can pass. At this time, the ant occurs in every county in South Texas in which the tortoise occurs except Starr and Zapata Counties and part of Jim Hogg County. However, the ant's distribution is spotty in most of these counties and is centered around water sources, in towns, or on farms and ranches where supplemental feeding of hay to livestock occurs. There are native fire ants over much of the range of the tortoise, but they are less aggressive and occur at vastly lower densities; their role as predators of compromised vertebrates is unknown but is probably minimal.

Protecting Texas tortoises seems to lead to no negative impacts to other species. In fact, habitat set aside for protecting ocelots (*Felis pardalis*) and jaguarundis (*Felis yagouaroundi*) provides protection for tortoises, but in Texas the amount of shared habitat is minimal. And, where tortoises and semilarge cats are coinhabitants, not a lot of thought has gone into whether the cats might have a negative impact on tortoises, because tortoise welfare falls well outside the veil of conscious care. However, southernmost Texas, especially the Lower Rio Grande Valley, is one of the fastest-developing areas in the United States, and this onslaught will continue. Urbanization is claiming significant tracts of land.

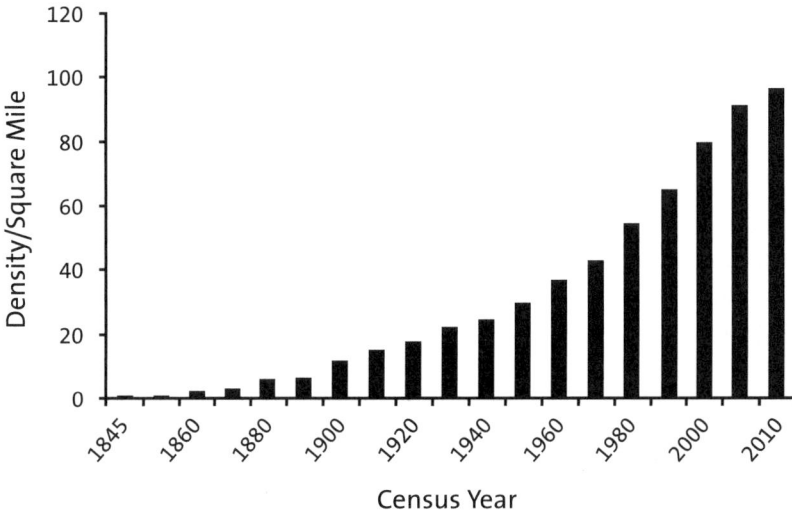

Figure 11.1. Human population density in Texas per square mile through 2010. More people mean more vehicles on roads, and a greater probability of anthropogenic disasters that will impact Texas tortoises negatively.

Most of the urbanization follows a progression: land is first cleared for agriculture, and agricultural land is subsequently converted to urban use. Relatively little shrubland is cleared initially for residential development. Lands along lake margins, resacas, the Rio Grande, the Arroyo Colorado, and the Laguna Madre were initially cleared for urban development.

The general progression leading to local extirpation of tortoises is a vortex of factors that sequentially fuse. Habitat destruction occurs in tandem with human population increases, leading to further negative environmental alterations that are followed by an increase in nonnative species, including subsidized predators, and increased removal of tortoises as food or for commercial purposes, notably the pet trade. While you can make laws against specific actions that negatively impact wildlife, you cannot make laws against cultural norms that encourage the abuses. And, when the issues of habitat destruction and species extinction peak, the bottom line is that unchecked human population growth is the common denominator. While a pittance is actually dedicated to protecting species worldwide, pharmaceutical companies and medical researchers spend billions of dollars annually to increase human fertility. More humans (fig. 11.1), more resources needed, more destruction of the ecological realm, and decreased sensitivity to the plight of our charges lead to more misery. This Malthusian trap was sprung on numerous dominant civilizations but nowhere with more fervor than in China during the Qing Dynasty (1644–1911) (Mann 2011). Perhaps George Santayana was correct: those who cannot remember the past are condemned to repeat it. And, we may have already entered "The Purse-Seine" (Jeffers 1937).

Status of Extant Populations

Texas

There is no doubt that populations of Texas tortoises have dwindled or disappeared in many places. The development of orchards and farms in once-prime habitat along vast stretches in the Lower Rio Grande Valley, the clearing of brush for cattle, and the establishment of commercial centers have all taken a toll on tortoise numbers. The tortoises have to contend not only with their natural predators but with new ones, such as feral hogs and domestic dogs. But by and large, human activities associated with land conversion and road traffic produce major negative impacts. Because of the extent of the tortoise's range, it is protected in vast areas of ranchland that are not likely to be developed. However, general work traffic within these ranches probably accounts for significant numbers of annual deaths. Anecdotal accounts lead us to believe that tortoise densities are vastly reduced from what they were in the 1960s and 1970s. In part, this impression might be a reflection of the significant mileage of high game fences that are in place along roads over much of the tortoise's range. These fences limit tortoise movements and increase mortality by trapping individuals and enhancing searches by predators (Engeman et al. 2004). Densities are generally low over much of the tortoise's range, but as long as large blocks of ranchland are in place, viable populations can be maintained.

Mexico

Little is known about the status of Texas tortoises in Mexico. There are no prohibitions there against their capture or use. We know of no sites where densities approach those attained in South Texas, and the low densities and harsh environment in Mexico may contribute to a lack of long-term studies there. Generally, what we know about this tortoise in Mexico is derived from distribution records (Bogert and Oliver 1945), unpublished field notes, and museum specimens. Although there are large gaps in our knowledge about the tortoise in different habitats in Texas, and these need to be filled, information about tortoise natural history in Mexico is sorely needed. This would be a challenging but productive enterprise. If the social climate in Mexico at this writing continues, it could impede investigations there for many years. After all, there is no need to fill one's head with data if one has no head.

Management

There are inherent problems in managing wide-ranging, long-lived species. We posit that every little bit helps, but wide-scale management is not feasible in South Texas. The Lower Rio Grande Valley Wildlife Corridor is destined to connect the existing natural tracts of brushland left along the lower stretch of the Rio Grande from Falcon Lake to the mouth of the river. The goal is to have 53,621 ha (132,500 ac) under the control of the Texas Parks and Wildlife Department (TPWD), the National Audubon Society, The Nature Conservancy, private landowners, and the Santa Ana and Laguna Atascosa National Wildlife Refuges. Texas tortoises should benefit greatly from this cooperative endeavor.

Several management-oriented studies are called for. The fate of long-term captives released into natural environments should be evaluated, possibly with radiotelemetry. At the same time, researchers could evaluate site-specific behavior. A detailed study of the potential for pathogen transfer should be conducted and a survey of known veterinarians who handle reptilian cases should be undertaken to get a handle on upper respiratory tract diseases. There is no ethical reason to release captive tortoises back into a natural environment if we cannot certify that they are disease free. Aside from that, captive-reared tortoises released into the environment could quite possibly be immunocompromised, having not been exposed to natural pathogens early in life.

Wardens are loath to confiscate captive Texas tortoises because they have no way of adequately handling them. This is especially trying considering the number of captive tortoises some people maintain. Policy should dictate how to provide those in the vanguard with support, but there is little reason to sit up and take notice if the outcome is to continue sitting.

Legal and Moral Issues

Writing a section on the care of captive tortoises has implications that are not lost on us. On the one hand we emphasize that the Texas tortoise should not be maintained out of the wild, and on the other hand we give suggestions for its care in captivity. Because there are so many tortoises known to be in captivity, and programs from TPWD that we hope will come to pass, we hope that more tortoises will be harbored legally. In addition, more tortoises will be transferred to rehabilitators; thus, we offer the information on care to help shorten the learning curve for those maintaining these tortoises.

It is illegal in Texas to possess a Texas tortoise without an appropriate permit, granted by the Texas Parks and Wildlife Department. Indeed, the TPWD lists the tortoise as threatened. However, many are in captivity and they are frequently maintained in a single-species group and mixed with additional individuals periodically or mixed with other turtle species, including exotic turtles. In one possible scenario, 18–30 tortoises, each with a potential life span of 70 years, will generally flourish in a backyard until property owners tire of their digging and their elimination of grass, or until the captors face their inevitable inability to care for the tortoises. When this time comes, the tortoises might be released into the wild, carrying with them pathogens accrued over the years to naive populations.

It is unwise to maintain tortoises with dogs. We have seen many tortoises with severe shell damage, sometimes with punctures through the carapace, because they were left alone with dogs that perceived the tortoises as animated chew-toys. Young, thin-shelled tortoises are extremely susceptible to injury. Strangely, while the initial reaction of a tortoise to harassment is to withdraw its head and limbs into its shell, if agitation continues the animal will extend these very appendages, with catastrophic consequences.

Tortoise injuries and diseases are expensive to treat professionally, but those who accept the responsibility of care and maintenance of a tortoise have an obligation to ensure proper veterinary care when the need arises. In addition, it is incumbent on those maintaining tortoises to educate themselves to ensure that there is a reasonable expectation that the tortoises will thrive in captivity. As humans, we surmise that we ourselves do better as couples, and we frequently transfer this line of reasoning to our pets. A person with a female tortoise may soon be on the lookout for a mate, a coinhabitant, a friend—another victim of circumstance to mitigate the perceived loneliness. Nothing could be further from the reality of the tortoise's needs, because Texas tortoises are solitary and have significant contact with the opposite sex only at mating time. More captive individuals require more food, food supplements, care during inclement weather (especially winter), and supplemental care when the caregiver is not present. If you see a tortoise on the roadway, **move it to safety and drive away as quickly as is legally allowed**. Feel good that you have helped another species, as precious as your own!

Pet Trade

It is illegal to barter Texas tortoises; however, it is also illegal to run stop signs. Laws and regulations governing tortoises are probably more helpful in reducing the number sold or traded than if no laws or regulations were in effect. Many Texas tortoises were imported into California from northeast Mexico via New Mexico to avoid Texas laws (Brame and Peerson 1969). Auffenberg and Weaver (1969) reported that 4,000 Texas tortoises were sent in one shipment to California, and Luckenbach (1982) cited a personal communication from Glen R. Stewart in 1974 that 8,000 tortoises were shipped from Mexico to California in two vans. In addition, Stewart estimated that 40,000 Texas tortoises were being imported into California each year. We estimate that in over 30 years of study, we have observed fewer than 800 individual tortoises. The gathering of thousands of Texas tortoises would be a daunting enterprise requiring a lot of manpower; thus, while we do not doubt that high numbers have been transported over the years, 4,000 to 40,000 individuals per year are unrealistic estimates.

Some people in South Texas maintain many tortoises in their backyards.

Thirty-two people who contacted us in 1994–95 maintained an average of 15 (1–72) Texas tortoises. These tortoises reproduce and are maintained with other native turtles, usually box turtles, and exotic species. Sick or predator-injured tortoises are expensive to treat and may be released back into the wild, where they can infect other tortoises.

One problem is the aging collector-maintainer who eventually realizes that the time is nigh to downsize and move into more easily maintained accommodations. What of the tortoises? What is their fate after being released back into the wild after years of pampered existence where the sharp edges of their physiological and ecological adaptations have been dulled by artificial care? The brief answer is that no one knows, nor, apparently, does anyone involved with tortoise protection really care to offer another option.

Captive Propagation

Currently there is no need to instigate captive propagation of Texas tortoises, followed by their release into natural habitats. There are enough large ranches south of San Antonio where tortoises are basically protected from most negative impacts, except vehicular traffic and predation. In addition, large blocks of prime habitat are now protected because they are associated with the Lower Rio Grande Valley Wildlife Corridor project, the U.S. Fish and Wildlife Service refuges, and Texas state parks.

General Care and Maintenance

Texas tortoises are known to spend considerable time walking the edges of enclosures. If they ever find an escape site, upon recapture they will return to that site time and again. Also, any such enclosure should also be an exclosure against animals that might injure or kill tortoises. Shade in the form of vegetation and a protective shelter that will allow them to escape extreme environmental conditions is imperative. In warmer climates, minimal covering with vegetation may be adequate, but as temperatures decline below 4.4°C (40°F), additional efforts are required. If the temperatures drop to 3.9°C (39°F), serious thought should be given to keeping the tortoises in a protective structure. This can be done by placing the animals in a container with wood shavings (preferably hardwood) and placing the container in an enclosed shelter. Depending on the severity of the winter temperatures, the box should not rest directly on a concrete floor but should be placed on a flattened cardboard box or its equivalent. How tortoises are to be protected in the winter will, of course, be determined by the number with which one has to deal.

In South Texas, captive tortoises are usually left to their own devices during the winter. Most do quite well with minimal cover, but remember that as temperature decreases tortoises lose their ability to alter their status, that is, to dig or shuffle deeper into the substrate. Sustained temperatures below 0°C (32°F) may freeze the spinal cord if the carapace is exposed, permanently paralyzing the hind limbs.

Texas tortoises can climb wire for short distances and have extraordinary resolve in attempting escape, especially in small enclosures. If the enclosure wall is less than 48 cm (18 in) high, then the corners should be rounded or a board tacked in place across the junction to deter escape. Tortoises tend to congregate in corners and have no compunction about scaling their comrades to facilitate escape. As one enthusiast cogently exclaimed, "They can climb like goats."

If there is a probability of predator activity, some tortoise aficionados recommend housing tortoises in secure quarters (see plate 26). We found that a safe harbor made of 4" × 4" cedar posts (DO NOT use treated posts) with a small port for exit and entry suffices, as the tortoises readily retire into it at night, and it is sufficient to protect animals during the winter if the roof is constructed with a layer of Styrofoam and Hardie Board. Several people we know have sheds in their backyards and tortoises simply come and go under them.

Captive adult tortoises rarely drink if their food has sufficient moisture content. Young tortoises, however, drink occasionally during hot weather. For some unknown reason, most Texas tortoises do not drink from a water-filled container. Drinking from a pan can be enhanced if a drip system is employed and the water is shallow. Remember that they drink through their nose, so the water does not have to be deep. Texas tortoises that attempt to drink with an open mouth probably have some underlying health problem.

If the net production of grass in the enclosure exceeds its consumption by tortoises, little supplemental feeding is necessary. Grass is helpful in maintaining dietary bulk and provides nutrition. Sporadic supplemental offerings of scraps such as cantaloupe, watermelon, squash, broccoli, cabbage, cauliflower, and tomatoes are helpful. Romaine lettuce is the only single food that we know of that meets most of the tortoises' dietary needs for maintenance and growth. On a diet solely of romaine lettuce, young tortoises grow rapidly, maintain good water balance, and appear normal in every way except that they eat a lot, perhaps three times what one would expect. Calcium supplements for reptiles as touted by pet stores are probably helpful to young animals. Avoid diets high in protein (cat or dog food). This does not mean the tortoises will not eat such foods, for they surely will. It means that the tortoises do not have the proper metabolic machinery to handle such foods on a routine basis and will eventually suffer. Avoid tough, old cactus cladophylls, as they contain (along with such vegetables as spinach) excessive amounts of oxalic acid, which may negatively impact kidney and neurological function.

Early on it is convenient and helpful to shift tortoises to a commercial tortoise food. We prefer Mazuri tortoise chow, but it should be supplemented with plenty of greens and fruit. The pellets are too hard for tortoises to crush efficiently but if pellets are soaked for 60 seconds in water, they are readily consumed, providing nutrition and fiber. Care should be exercised not to oversoak the pellets, as they will disintegrate. Tortoises require a certain amount of abrasive food to maintain the horny sheaths covering their jaws. Distortion of these sheaths induced from eating soft food is difficult and expensive to correct.

Tortoises do drown. On occasion, we have heard about people picking them up and, not knowing anything about tortoises and assuming they are aquatic, placing them in a container of water where they can swim. Eventually they will tire and drown. Do not assume that the animal is beyond help if it is flaccid and unresponsive, as tortoises have tremendous resilience. Water entering the circulatory system from the lungs initiates all manner of physiological havoc from blood cell destruction to ventricular fibrillation, so it is important to reduce the amount of this water as soon as possible. Hold the animal upside down and open its mouth. Pump the hind legs vigorously to drain water from the lungs and stomach. Next, close off the mouth with your fingers and blow air into the tortoise's nose to inflate the lungs. Do this about 15 times per minute for no more than 20 minutes. After this time, if the tortoise has no visible change in muscle activity, that is, no retraction of the legs and head when physically extended, it is not a good sign. If the tortoise responds to having its head extended but the neck-head plane is twisted rather than straight, the tortoise has suffered neural damage from which it will probably not fully recover. In the event that the tortoise recovers, it will need warmth and a good dose of antibiotics because the water will have transmitted microorganisms into its inner confines. If the tortoise does not eat, it should be tube- or force-fed daily for five days. A mixture of warm baby food including carrots, squash, and some fruit works well. After 10 days, if the tortoise has not shown significant progress and is lethargic, and its legs dangle when it is picked up, humane termination should be considered. See your veterinarian at this time.

Care of Tortoise Eggs

Remember that Texas tortoises are inhabitants of semiarid shrublands and their eggs do not require water to develop. If you discover an egg, mark the top with a pen or pencil and try to keep the mark in the same position when moving the egg. We know some people who have hatched many eggs by simply placing them in a flowerpot, with no resting medium, and placing the pot in a shaded spot. We found the following to work rather well.

Place moist masonry sand into a tortilla warmer to a depth of about 2.5 cm (1 in). (CAUTION: Do not use vermiculite, as it may come with a good dose of asbestos.) Nestle the egg into the sand such that about half the egg is surrounded. Mark the date on the container or eggs and replace the cover. Place the container in a warm, secure place and remember that temperature determines hatching time and the sex of the offspring. Check for hatching in about 88–118 days. Do not touch the eggs; they will either hatch or they will not, and nothing will be gained by attempting to hurry the process. It will probably take several days for pipping to be complete, but the tortoise should exit the shell of its own accord. It will appear misshapen, and you will wonder how such a large creature was housed in so small an egg. Envision your thumb touching the tip of your middle finger; that is how the tortoise develops within its eggshell. It will take several days for the fold to flatten, and for the yolk to reabsorb. You can admire computers, airplanes, steam and gasoline engines, or the Seven Wonders of the World, but when the rubber meets the road, the cleidoic egg and its developmental product stand as one of the major evolutionary accomplishments in the grand procession of life on earth. Flight by vertebrates was a novelty; the cleidoic egg was a fundamental world-changing development!

Hatchling tortoises seek out dark places immediately upon hatching, and they may become reclusive. They will eat the tender edges of romaine lettuce within a couple of days and then be well on their way. On occasion, every week or two, place the tortoise in a container of water such that its depth is no more than half the height of the tortoise. Once you have reassured yourself that the water depth is adequate, leave the tortoise for 10–20 minutes. During this time it should void its cloacal contents.

If you maintain young tortoises inside during the winter, they can be kept in a terrarium (no free water) with a heating stone at one end. Tightly packed, moistened organic mulch works well because it develops a hard covering (CAUTION: do not use mulch with chemical enhancers such as fertilizer). There should be a dark, cavernous retreat. Remember that these tortoises are thigmotaxic and will tend to push up and into or under a covering. Whatever the substrate, make sure that the tortoises do not inadvertently eat it, because they may develop an intestinal block. Food should be presented on a clean surface, such as a piece of cardboard. As soon as possible, move the tortoises to an outside enclosure where they will have access to sunlight and shade as well as to soil microorganisms.

Captive tortoises should be inspected every month or so to check for infections and injuries. In addition, clipping the nails to prevent distortion of the digits will be necessary. A general antiseptic that seems to work well for tortoises can be made by mixing Betadine solution and 3% peroxide (50:50). This can be squirted into rather tight recesses of the tortoise.

Veterinary Care

Not all veterinarians are professionally or personally trained to service reptiles. All veterinarians that we have worked with were certainly helpful and made every effort to accommodate tortoises; however, a veterinarian with specific practical expertise in reptiles is more apt to be able to provide efficient diagnosis and will have data on anesthesia and treatment regimens for conditions many other veterinarians have not experienced. Reptilian kidneys do not handle drugs in the manner of those of mammals or birds. Therefore, special knowledge of reptile treatment regimens is desirable. A captive tortoise in your possession is in no way less entitled to proper medical care than a warm-blooded, fuzzy, and charismatic pet. Caring for a tortoise is like dancing with bears: you might not be able to quit just because you are tired! Veterinary care is expensive, but if the tortoise is in your possession it is your responsibility to see that it is cared for properly.

Initially it may be difficult to ascertain whether a tortoise is merely basking or is expressing symptoms of illness. If it is basking, it will begin normal behavior immediately upon being disturbed. If a tortoise remains lethargic when disturbed, especially if both forelimbs are extended equally, the tops of the forelimbs lie flat on the resting surface (see fig. 3.1), the eyelids remain closed and swollen, and the head is extended, then the tortoise needs evaluation. If the animal exudes mucous bubbles from the mouth or nasal openings, immediate veterinary care is necessary. In the absence of bubbles, Baytril (Enrofloxacin) was the antibiotic of choice for tortoises for many years, but alas, resistant bacteria in some reptiles caught up with its effectiveness. In addition, Baytril had certain iatrogenic effects, and tortoises responded to the intramuscular injections as though they were painful. Baytril is also known, on rare occasions, to induce necrosis, leading to possible loss of the injected limb. We suggest giving the Baytril orally at a rate of 10–20 mg/kg for 7–10 days. Follow each dose with 10 cc of water to ease the negative taste sensation. Fortaz (Ceftazidime) is currently getting the job done, and tortoises seem to tolerate it well. It is injected subcutaneously at a rate of 20 mg/kg every 72 hours. Because reptiles do not process antibiotics as mammals do, it is important to hydrate the tortoise to protect kidney function. Never treat tortoises with antibiotics without consulting a veterinarian.

Separate Maintenance from Other Tortoises

Many exotic tortoises carry heavy parasite loads, especially roundworms. Typically, millions of eggs are extruded with the feces and mingle with the soil and vegetation. When an uninfected tortoise swallows these eggs, parasite hatching and maturation ensue. The more crowded the tortoises, the more apt a severe infection is to develop. Remember, even if a tortoise does not show outward signs of infection, it may still be heavily infested with parasites. Treatment and elimination of roundworms is rather simple if the eggs have not passed into the maintenance enclosure. Therefore, never mingle tortoises (including box turtles) of unknown health status. Pyrantel pamoate is an excellent dewormer for tortoises but must be given at the appropriate dosage and in concert with veterinary care.

12

The Future

Species Significance and Bearing on Ecological Problems

> We of an older generation can get along with what we have, though with growing hardship; but in your full manhood and womanhood you will want what nature once so bountifully supplied and man so thoughtlessly destroyed; and because of that want you will reproach us, not for what we have used, but for what we have wasted.
>
> —Theodore Roosevelt, an Arbor Day message to the schoolchildren of the United States, April 15, 1907

Tortoises are remnants of an ancient lineage. They date back to the Eocene, and extant tortoises are now found on all continents except Antarctica. However, only five living species remain in North America. Texas tortoises are the smallest and most sexually dimorphic of these five species (Auffenberg and Weaver 1969; Rose and Judd 1982). Survival of the Texas tortoise is threatened by habitat reduction and other human-caused perturbations (Bury 1982; Bury and Germano 1994), but it escapes dire straits because large blocks of ranchland, virtually off limits to the public, provide safe haven for many populations.

Alas, the Texas tortoise is not a federally protected species. The Texas Parks and Wildlife Department considers the Texas tortoise a threatened species, but enforcement officers are not prone to take action. An enforcement officer confiscating a tortoise is faced with a dilemma: he or she cannot release the tortoise, and there are few options of places to send it. Zoological parks generally have more Texas tortoises than they wish to have, and zoo personnel are reluctant to accept animals because of the possible transmission of *Mycoplasma* (Judd and Rose 2000). It would be useful if the federal government would recognize the Texas tortoise as a threatened species, as it would increase awareness of the status of the species and help with the enforcement of laws to protect it. It is best to conserve water when you have lots of it, and not wait until the middle of a drought!

Much of the basic biology of the Texas tortoise remains unknown. A person familiar with the literature could easily identify 25 or more topics in need of research in a matter of minutes. Throughout this book, we have identified areas where information is lacking and areas where controversy exists. Indeed, this was a major objective of the book. Next, we suggest a few topics that are most deserving of future study.

The distribution of Texas tortoises in Mexico is poorly known and we have no information on other aspects of the biology of the species in this large portion (over half) of its geographic range (Rose and Judd 1982; Rose and Judd 1989; Germano and Bury 1994). Basic life history information on tortoise populations in Mexico is sorely needed for the planning and implementation of conservation measures. Information on Mexican populations is especially important because rangelands there are rapidly being converted to agricultural fields. For example, in 1953–54 the total area devoted to agricultural production in Tamaulipas, Mexico, was 243,800 ha (602,186 ac), and in 1980–81, 1,310,000 ha (3,235,700 ac) were devoted to agriculture (Jarhrsdoerfer and Leslie 1988).

The Texas tortoise is the poster child of the Tamaulipan Biotic Province. Its geographic range and the boundary of the Tamaulipan Biotic Province are virtually the same (map 3). This close correspondence of the two makes a strong case for conserving the species by conserving its habitat. Thus, if one wishes to conserve Texas tortoises, efforts should focus on conservation of the thorn-shrub and grassland communities of the Tamaulipan Biotic Province. The principal communities in South Texas, identified by McLendon (1991), that we think are likely to support tortoises are: little bluestem–trichloris (*Schizachyrium scoparium–Chloris pluriflora*) community, seacoast bluestem–balsamscale (*Schizachyrium littorale–Elionurus tripsacoides*) community, live oak–post oak (*Quercus virginiana–Quercus stellata*) community, mesquite–granjeno (*Prosopis glandulosa–Celtis pallida*) community, huisache–prickly pear (*Acacia minuata–Opuntia engelmannii*) community, mesquite–prickly pear (*Prosopis glandulosa–Opuntia engelmannii*) community, guajillo-cenizo (*Acacia berlandieri–Leucophyllum frutescens*) community, blackbrush–twisted acacia (*Acacia rigidula–Acacia schaffneri*) community, and creosote–prickly pear (*Larrea divaricata–Opuntia engelmannii*) community. Studies by Auffenberg and Weaver (1969), Judd and Rose (1983, 1989), Rose and Judd (1975), and Bury and Smith (1986) have all been in mesquite–prickly pear communities or mesquite-granjeno communities of Cameron and Willacy Counties. The studies of Hellgren et al. (2000) and Kazmaier et al. (2001a, 2001b, 2001c) were in the mesquite-acacia community of Dimmit and La Salle Counties. Consequently, most of the communities listed above have not been surveyed and tortoise abundance in these communities is unknown. Determination of tortoise abundance in all the communities where the species occurs in Texas and Mexico should be a major future focus. In addition, efforts should be made to delineate the geographic range of the tortoise in the southeastern portion of its range in Mexico.

Life history and ecological studies of long-lived species require commitment, energy, and resources (Wilbur 1975; Tinkle et al. 1981), but we are seeing more being done. Texas tortoises are an excellent example of a long-lived organism, and they possess many attributes that make them an excellent model for experimental field studies. They are terrestrial, active during the day, and large enough to see at a distance. They are slow moving, easy to catch, and easy to mark by etching a number on the carapace with a Dremel tool. Transmitters and passive integrated transponders (PIT tags) can be attached or easily inserted into the forearm. Densities are frequently high enough that sufficient numbers are available for statistical analyses. Despite these practical advantages for study, much remains to be learned about the basic biology of the species.

Most research on life histories has focused on short-lived species in large part because such a focus is more practical for researchers attempting to produce publications. Usually, one or more generations can be studied in a year and publications resulting from that study can be in print in another year or two. In contrast, studying long-lived species requires monitoring individually marked animals over many years. This is expensive and hard to do, and the return for such effort is small when measured in terms of the number of publications produced per unit of study time. In addition, it is difficult to maintain consistency in the timing of survey periods, a point not missed by journal review editors.

Bury and Germano (1994) stated that North American tortoises are *keystone species*. These are species that have strong effects on community structure even though they may not be particularly abundant. Few terrestrial communities are thought to be organized by keystone species (Krebs 2001). Determining if a species is a keystone species requires obtaining a detailed understanding of the food web structure of a community by experimental analysis (Krebs 2001). Certainly such information is lacking for Texas tortoises. Indeed, we have only a rudimentary knowledge of the food habits of the species, and there have been no experimental manipulations such as removal of the tortoise in any given community. Obtaining detailed information on the kinds and quantities of foods consumed by Texas tortoises in different plant communities should be a high priority for future studies (see Vecchio et al. 2011 for a good model). The Texas tortoise may be a keystone species in its communities, but it is premature to make such a claim. Nevertheless, we hope that it is true!

Comparison of data for coastal populations (Auffenberg and Weaver 1969; Rose and Judd 1982; Judd and Rose 1983; Bury and Smith 1986; Judd and Rose 1989; Rose and Judd 1989) and inland populations (Hellgren et al. 2000; Kazmaier et al. 2001a, 2001b, 2001c) shows that there is geographic variation in size, density, and reproductive parameters, but patterns of variability in relation to climatic factors such as temperature and rainfall are unknown, as the existing studies have focused on just two areas in Texas. Information is needed on the biology of populations along a north-south axis through the species' geographic range, as this would extend from the South Temperate Zone through the sub-tropics to the tropics and would permit the correlation of biological aspects with temperature. Similar information is needed along an axis of increasing aridity throughout the geographic range of the species.

References

Agassiz, L. 1857. *Contributions to the Natural History of the United States of America.* Boston: Little, Brown.

Agha, M., J. E. Lovich, J. R. Ennen, and E. Wilcox. 2013. "Nest-guarding by female Agassiz's desert tortoise (*Gopherus agassizii*) at a wind-energy facility near Palm Springs, California." *Southwestern Naturalist* 58:254–57.

Allender, M., M. Abd-Eldaim, J. Schumacher, D. McRuer, L. Christian, and M. Kennedy. 2011. "PCR prevalence of *Ranavirus* in free-ranging eastern box turtles (*Terrapene carolina carolina*) at rehabilitation centers in three southeastern US states." *Journal of Wildlife Diseases* 47:759–64.

Alvarez, T. 1963. "The recent mammals of Tamaulipas, México." *University of Kansas Publications Museum Natural History* 14:363–473.

Arata, A. A. 1958. "Notes on the eggs and young of *Gopherus polyphemus* (Daudin)." *Quarterly Journal of the Florida Academy of Science* 21:274–80.

Auffenberg, W. 1969. *Tortoise Behavior and Survival.* Chicago: Rand McNally.

———. 1976. "The genus *Gopherus* (Testudinidae). I. Osteology and relationships to extant species." *Bulletin of the Florida State Museum* 20:47–110.

Auffenberg, W., and R. Franz. 1978a. "*Gopherus.*" *Catalogue of American Amphibians and Reptiles* 211.1–211.2.

———. 1978b. "*Gopherus berlandieri.*" *Catalogue of American Amphibians and Reptiles* 213.1–213.2.

Auffenberg, W., and J. B. Iverson. 1979. "Demography of terrestrial turtles." In *Turtles: Perspectives and Research*, edited by H. Morlock and M. Harless, 541–69. New York: John Wiley & Sons.

Auffenberg, W., and W. G. Weaver. 1969. "*Gopherus berlandieri* in southeastern Texas." *Bulletin of the Florida State Museum* 13:141–203.

Averill-Murray, R. C. 2011. "Comments on the status of the desert tortoise(s)." *Herpetological Review* 42:500–501.

Barton, S. L. 2006. "Shell damage." In *Reptile Medicine and Surgery*, edited by D. R. Madar, 893–99. St. Louis: Saunders.

Baze, W. B., and F. R. Horne. 1970. "Urogenesis in Chelonia." *Comparative Biochemistry and Physiology* 34:91–100.

Beasom, S. L. 1974. "Selectivity of predator control: Techniques in south Texas." *Journal of Wildlife Management* 38:837–44.

Beasom, S. L., and C. J. Scifres. 1977. "Population reactions of selected game species to aerial herbicide applications in south Texas." *Journal of Wildlife Management* 30:138–42.

Benton, A. H., and W. E. Werner, Jr. 1974. *Field Biology and Ecology*, 3rd ed. New York: McGraw Hill.

Berry, J. F., and R. Shine. 1980. "Sexual size dimorphism and sexual selection in turtles (Order Testudines)." *Oecologia* 44:185–91.

Berry, K. H., D. J. Morafka, and R. W. Murphy. 2002. "Defining the desert tortoise(s): Our first priority for a coherent conservation strategy." *Chelonian Conservation and Biology* 4:249–62.

Bethea, N. J. 1972. "Effects of temperature on heart rate and rates of cooling and warming in *Terrapene ornata*." *Comparative Biochemistry and Physiology* 41A:301–305.

Blair, W. F. 1950. "The biotic provinces of Texas." *Texas Journal of Science* 2:93–117.

Bogert, C. M., and J. A. Oliver. 1945. "A preliminary analysis of the herpetofauna of Sonora." *Bulletin of the American Museum of Natural History* 83:303–425.

Bour, R., and A. Dubois. 1984. "*Xerobates agassiz*, 1857, synonyme plus ancien de *Scaptochelys* Bramble, 1982 (Reptilia, Chelonii, Testudinidae)." *Bulletin Mensuel de la Société Linnéenne de Lyon* 53:30–32.

Bowen, G. S. 1977. "Prolonged western equine encephalitis viremia in the Texas tortoise (*Gopherus berlandieri*)." *American Journal of Tropical Medicine and Hygiene* 26:171–75.

Bramble, D. M. 1974. "Occurrence and significance of the os transiliens in gopher tortoises." *Copeia* 1974:102–109.

———. 1982. "*Scaptochelys*: Generic revision and evolution of gopher tortoises." *Copeia* 1982: 852–67.

Brame, A. H., Jr., and D. J. Peerson. 1969. "Tortoise ID." *International Turtle and Tortoise Society Journal* 3:8–12.

Braun, J. K., and M. A. Mares. 1989. "*Neotoma microceps*." *Mammalian Species* 330:1–9.

Brode, W. E. 1959. "Notes on behavior of *Gopherus polyphemus*." *Herpetologica* 15:101–102.

Brown, A. E. 1908. "Generic types of nearctic Reptilia and Amphibia." *Proceedings of the Academy of Natural Sciences of Philadelphia* 60:112–27.

Brown, B. C. 1950. "An annotated checklist of the reptiles and amphibians of Texas." Waco, Tex.: Baylor University Studies.

Brown, D. R., I. M. Schumacher, G. S. McLaughlin, L. D. Wendland, M. B. Brown, P. A. Klein, and E. R. Jacobson. 2002. "Application of diagnostic tests for mycoplasmal infections of desert and gopher tortoises, with management recommendations." *Chelonian Conservation and Biology* 4:497–507.

Brown, M. B., I. M. Schumacher, P. A. Klein, K. Harris, T. Corre, and E. R. Jacobson. 1994. "*Mycoplasma agassizii* causes upper respiratory tract disease in the desert tortoise." *Infection and Immunity* 62:4580–86.

Burk, V. J., J. W. Gibbons, and J. L. Greene. 1994. "Prolonged nesting forays by common mud turtles (*Kinosternon subrubrum*)." *American Midland Naturalist* 131:190–95.

Bury, R. B. 1982. "An overview." In *North American Tortoises: Conservation and Ecology*, edited by R. B. Bury, v–vii. U.S. Fish and Wildlife Service, Wildlife Research Report 12.

Bury, R. B., and D. J. Germano. 1994. "Biology of North American tortoises: Introduction." In *Biology of North American Tortoises*, edited by R. B. Bury and D. J. Germano, 1–6. Washington, D.C.: U.S. Dept. of the Interior, National Biological Survey, Fish and Wildlife Research 13.

Bury, R. B., and E. L. Smith, 1986. "Aspects of the ecology and the management of the tortoise *Gopherus berlandieri* at Laguna Atascosa, Texas." *Southwestern Naturalist* 31:387–94.

Cagle, F. R. 1950. "The life history of the slider turtle, *Pseudemys scripta troostii* (Holbrook)." *Ecological Monographs* 20:31–54.

Cagle, F. R., and J. Tihen. 1948. "Retention of eggs by the turtle *Deirochelys reticularia*." *Copeia* 1948:66.

Campbell, H. W., and W. E. Evans. 1967. "Sound production in two species of tortoises." *Herpetologica* 23:204–209.

———. 1972. "Observations on the vocal behavior in chelonians." *Herpetologica* 28:277–80.

Carr, A. 1952. *Handbook of Turtles.* Ithaca, N.Y.: Cornell University Press / Comstock Publishing.

Chelazzi, G., and G. Delfino. 1986. "A field test on the use of olfaction in homing by *Testudo hermanni* (Reptilia: Testudinidae)." *Journal of Herpetology* 20:451–55.

Cheylan, M. 1984. "The true status and future of Hermann's tortoise *Testudo hermanni robertmertensi* Wermuth 1952 in Western Europe." *Amphibia-Reptilia* 5:17–26.

Congdon, J. D., and J. W. Gibbons. 1985. "Egg components and reproductive characteristics of turtles: Relationships to body size." *Herpetologica* 41:194–205.

Cottam, C., W. C. Glazner, and G. G. Raun. 1959. "Notes on food of moccasins and rattlesnakes from Welder Wildlife Refuge, Sinton, Texas." *Contributions of the Welder Wildlife Foundation* 45:1–12.

Crews, D., and M. C. Moore. 1993. "Psychobiology of reproduction of unisexual whiptail lizards." In *Biology of Whiptail Lizards* (*Genus* Cnemidophorus), edited by J. W. Wright and L. J. Vitt. Norman: Oklahoma Museum of Natural History.

Crumley, C. R. 1994. "Phylogenetic systematics of North American tortoises (genus *Gopherus*): Evidence for their classification." In *Biology of North American Tortoises*, edited by R. B. Bury and D. J. Germano, 7–32. Washington, D.C.: U.S. Dept. of the Interior, National Biological Survey, Fish and Wildlife Research 13.

Crumley, C. R., and L. L. Grismer. 1994. "Validity of the tortoise *Xerobates lepidocephalus* Ottley and Velazques in Baja California." In *Biology of North American Tortoises*, edited by R. B. Bury and D. J. Germano, 33–37. Washington, D.C.: U.S. Dept. of the Interior, National Biological Survey, Fish and Wildlife Research 13.

Culbertson, G. 1907. "Some notes on the habits of the common box turtle (*Cistudo carolina*)." *Proceedings of the Indiana Academy of Science* 1907:78–79.

Davis, A. M. 1942. "A study of Boscaje de la Palma in Cameron County, Texas and of *Sabal texana*." MA thesis, University of Texas, Austin.

Dearing, M. D., A. M. Mangione, W. H. Karasov, S. Morzunov, E. Otteson, and S. St. Jeor. 1998. "Prevalence of hantavirus in four species of *Neotoma* from Arizona and Utah." *Journal of Mammalogy* 79:1254–59.

Diamond, D. D., D. H. Riskind, and S. L. Orzell. 1987. "A framework for plant community classification and conservation in Texas." *Texas Journal of Science* 39:203–21.

Diaz-Figueroa, O., and M. A. Mitchell. 2006. "Gastrointestinal anatomy and physiology." In *Reptile Medicine and Surgery*, edited by D. R. Madar, 893–99. St. Louis: Saunders.

Dice, L. R. 1943. *The Biotic Provinces of North America.* Ann Arbor: University of Michigan Press.

Diffie, S., J. Miller, and K. Murphy. 2010. "Laboratory observations of red imported fire ant (Hymenoptera: Formicidae) predation on reptilian and avian eggs." *Journal of Herpetology* 44:294–96.

Dixon, J. R. 2000. *Amphibians and Reptiles of Texas*, 2nd ed. College Station: Texas A&M University Press.

Dodd, C. K. 1995. "Disarticulation of turtle shells in north-central Florida: How long does a shell remain in the woods?" *American Midland Naturalist* 134:378–87.

———. 2001. *North American Box Turtles: A Natural History*. Norman: University of Oklahoma Press.

Douglas, J. 1986. "Patterns of mate-seeking and aggression in a southern Florida population of the gopher tortoise, *Gopherus polyphemus*." *Proceedings of the Desert Tortoise Council Symposium* 1986:155–99.

Duges, A. 1888. "La tortuga Polifemo." *La Naturaleza*, 2nd ser., 1:146–47.

Dunham, A. E., P. J. Morin, and H. M. Wilbur. 1988. "Methods in the study of reptile populations." In *Biology of the Reptilia*. Vol. 16, *Ecology*, edited by B. C. Gans and R. Huey, 331–86. New York: Academic Press.

Dupree, A. H. 1968. *Asa Gray*. New York: Atheneum.

Edgren, R. A. 1960. "Ovulation time in the musk turtle, *Sternotherus odoratus*." *Copeia* 1960:60–61.

Edwards, T., C. J. Jarchow, C. A. Jones, and K. E. Bonnie. 2010. "Tracing genetic lineages of captive desert tortoises in Arizona." *Journal of Wildlife Management* 74:801–807.

Eglis, A. 1962. "Tortoise behavior: A taxonomic adjunct." *Herpetologica* 18:1–8.

Emlen, S. T., and L. W. Oring. 1977. "Ecology, sexual selection, and evolution of mating systems." *Science* 197:215–23.

Engeman, R. M., M. J. Pipas, and H. T. Smith. 2004. "*Gopherus berlandieri* (Texas tortoise) mortality." *Herpetological Review* 35:54–55.

Ernst, C. H., and R. W. Barbour. 1972. *Turtles of the United States*. Lexington: University Press of Kentucky.

Everitt, J. H., and M. A. Alaniz. 1981. "Nutrient content of cactus and woody plant fruits eaten by birds and mammals in south Texas." *Southwestern Naturalist* 26:301–305.

Ewert, M. A. 1979. "The embryo and its egg: Development and natural history." In *Turtles: Perspectives and Research*, edited by M. Harless and H. Morlock, 333–413. New York: John Wiley & Sons.

Fields, J. R., T. R. Simpson, R. W. Manning, and F. L. Rose. 2003. "Food and selective foraging by the Texas river cooter (*Pseudemys texana*) in Spring Lake, Hays County, Texas." *Journal of Herpetology* 37 (4): 726–29.

Fitch, H. S., and M. V. Plummer. 1975. "A preliminary ecological study of the soft shelled turtle, *Trionyx muticus*, in the Kansas River." Israel Journal of Zoology 24:28–42.

Fritz, U., and O. R. P. Bininda-Emonds. 2007. "When genetics meets nomenclature: Tortoise phylogeny and the shifting generic concepts of *Testudo* and *Geochelone*." *Zoology* 110:298–307.

Fujii, A., and M. R. J. Forstner. 2010. "Genetic variation and population structure of the Texas tortoise, *Gopherus berlandieri* (Testudinidae), with implications for conservation." *Chelonian Conservation and Biology* 9:61–69.

Fulbright, T. E., and S. L. Beasom. 1987. Long-term effects of mechanical treatments on white-tailed deer browse. *Wildlife Society Bulletin* 15:560–64.

Galeotti, P., R. Sacchi, M. Fasola, D. P. Rosa, M. Marchesi, and D. Ballasina. 2005a. "Courtship displays and mounting calls are honest, condition-dependent signals that influence mounting success in Hermann's tortoises." *Canadian Journal of Zoology* 83:1306–13.

Galeotti, P., R. Sacchi, D. P. Rosa, and M. Fasola. 2005b. "Female preference for fast-rate, high-pitched calls in Hermann's tortoises *Testudo hermanni*." *Behavioral Ecology* 16:301–308.

Galeotti, P., R. Sacchi, R. D. Pellitteri, and M. Fasola. 2007. "Olfactory discrimination of species, sex and sexual maturity by the Hermann's tortoise *Testudo hermanni."* *Copeia* 2007:980–85.

Geiser, S. W. 1948. *Naturalist of the Frontier,* 2nd ed. Dallas: Southern Methodist University Press.

Germano, D. J. 1994. "Comparative life histories of North American tortoises." In *Biology of North American Tortoises,* edited by R. B. Bury and D. J. Germano, 175–85. Washington, D.C.: U.S. Dept. of the Interior, National Biological Survey, Fish and Wildlife Research 13.

Germano, D. J., and R. B. Bury. 1994. "Research on North American tortoises: A critique with suggestions for the future." In *Biology of North American Tortoises,* edited by R. B. Bury and D. J. Germano, 187–204. Washington, D.C.: U.S. Dept. of the Interior, National Biological Survey, Fish and Wildlife Research 13.

Gibbons, J. W. 1968. "Reproductive potential, activity, and cycles in the painted turtle, *Chrysemys picta."* *Ecology* 49:399–409.

———. 1982. "Reproductive patterns in freshwater turtles." *Herpetologica* 38:222–27.

Gibbons, J. W., J. L. Greene, and K. K. Patterson. 1982. "Variation in reproductive characteristics of aquatic turtles." *Copeia* 1982:776–84.

Gienger, C. M., and C. R. Tracy. 2008. Ecological interactions between Gila monsters (*Heloderma suspectum*) and desert tortoises (*Gopherus agassizii*)." *Southwestern Naturalist* 53:265–68.

Gist, D. H., and J. M. Jones. 1989. "Sperm storage within the oviduct of turtles." *Journal of Morphology* 199:379–84.

Goff, M. L., and F. W. Judd. 1981. "The first record of a chigger from the Texas tortoise, *Gopherus berlandieri."* *Southwestern Naturalist* 26:83–84.

Gould, F. W. 1969. *Texas Plants: A Checklist and Ecological Summary.* Texas Agricultural Experiment Station MP-585. College Station: Texas A&M University.

Grafen, A. 1988. "On the uses of data on lifetime reproductive success." In *Studies of Individual Variation in Contrasting Breeding Systems,* edited by T. H. Clutton-Brock, 454–71. Chicago: University of Chicago Press.

Grant, C. 1936. "The southwestern desert tortoise, *Gopherus agassizi."* *Zoologica* (NY) 21:225–29.

———. 1960. "Differentiation of the southwestern tortoises (genus *Gopherus*), with notes on their habits." *Transactions of the San Diego Society of Natural History* 12:441–48.

Gulick, W. L., and H. Zwick. 1966. "Auditory sensitivity of the turtle." *Psychological Research* 16:47.

Gunter, G. 1945. "The northern range of Berlandier's turtle." *Copeia* 1945:175.

Hamilton, R. D. 1944. "Notes on mating and migration in Berlandier's turtle." *Copeia* 1944:62.

Hellgren, E. C., R. T. Kazmaier, D. C. Ruthven, III, and D. R. Synatzske. 2000. "Variation in tortoise life history: Demography of *Gopherus berlandieri."* *Ecology* 81:1297–310.

Herbert, J. M., R. W. Dixon, and J. L. Isom. 2005. "A tropical weather vulnerability assessment for Texas coastal counties." *Texas Journal of Science* 57:187–96.

Householder, V. 1950. "Courtship and coition of the desert tortoise." *Herpetologica* 6:11.

Hutchison, V. H., A. Vinegar, and R. J. Kosh. 1966. "Critical thermal maxima in turtles." *Herpetologica* 11:32–41.

Huxley, J. S., and G. Tessier. 1936. "Terminology of relative growth." *Nature* 137:780–81.

Inglis, J. M., B. A. Brown, C. A. McMahan, and R. E. Hood. 1986. "Deer-brush relationships on the Rio Grande Plain, Texas." Texas Agricultural Experiment Station Contribution No. TA 16129.

Iverson, J. B. 1980. "The reproductive biology of *Gopherus polyphemus* (Chelonia: Testudinidae)." *American Midland Naturalist* 103:353–59.

———. 1990. "Nesting and parental care in the turtle, *Kinosternon flavescens*." *Canadian Journal of Zoology* 68:230–33.

Jackson, C. G., Jr., J. A. Trotter, T. H. Trotter, and M. W. Trotter. 1976. "Accelerated growth rate and early maturity in *Gopherus agassizii* (Reptilia: Testudines)." *Herpetologica* 32:139–45.

Jacobson, E. R., J. M. Gaskin, M. B. Brown, R. K. Harris, C. H. Gardiner, J. L. LaPointe, H. P. Adams, and C. Reggiardo. 1991. "Chronic respiratory tract disease of free-ranging desert tortoises, *Xerobates agassizii*." *Journal of Wildlife Diseases* 27:296–316.

Jacobson, E. R., T. J. Wronski, J. Schumacher, C. Reggiardo, and K. Berry. 1994. "Cutaneous dyskeratosis in free-ranging desert tortoises, *Gopherus agassizii*, in the Colorado Desert of southern California." *Journal of Zoo and Wildlife Medicine* 25:68–81.

Jarhrsdoerfer, S. E., and D. M. Leslie, Jr. 1988. "Tamaulipan brushland of the Lower Rio Grande Valley of South Texas: Description, human impacts, and management options." U.S. Fish and Wildlife Service, Biological Report 88 (36).

Jeffers, R. 1937. "The Purse-Seine." In *Such Counsels You Gave to Me and Other Poems*. New York: Random House.

Jennings, R. D. 1985. "Biochemical variation of the desert tortoise, *Gopherus agassizii*." Unpublished MS thesis, University of New Mexico.

Johnson, A. J., A. P. Pessier, J. F. X. Wellehan, R. Brown, and E. R. Jacobson. 2005. "Identification of a novel herpesvirus from a California desert tortoise (*Gopherus agassizii*)." *Veterinary Microbiology* 111:107–116.

Johnson, A. J., A. P. Pessier, J. F. X. Wellehan, A. Childress, T. M. Morton, N. L. Stedman, D. C. Bloom, W. Belzer, V. R. Titus, R. Wagner, J. W. Brooks, J. Spratt, and E. R. Jacobson. 2008. "*Ranavirus* infection of free-ranging and captive box turtles and tortoises in the United States." *Journal of Wildlife Diseases* 44:851–63.

Johnson, M. J., C. Guyer, S. M. Hermann, J. Eubanks, and W. K. Michener. 2009. "Patterns of dispersion and burrow use support scramble competition polygyny in *Gopherus polyphemus*." *Herpetologica* 65:214–18.

Johnston, M. C. 1963. "Past and present grasslands of southern Texas and northeastern Mexico." *Ecology* 44:456–66.

Jones, F. W. 1915. "The chelonian type of genitalia." *Journal of Anatomical Physiology* 10:393–406.

Judd, F. W. "Tamaulipan Biotic Province." In *The Laguna Madre of Texas and Tamaulipas*, edited by J. W. Tunnell, Jr., and F. W. Judd, 38–58. College Station: Texas A&M University Press.

———. 2004. "Community ecology of freshwater, brackish and salt marshes of the Rio Grande delta." *Texas Journal of Science* 56:103–122.

Judd, F. W., and J. C. McQueen. 1980. "Incubation, hatching, and growth of the tortoise, *Gopherus berlandieri*." *Journal of Herpetology* 14:377–80.

———. 1982. "Notes on the longevity of *Gopherus berlandieri* (Testudinidae)." *Southwestern Naturalist* 27:230–32.

Judd, F. W., and F. L. Rose. 1977. "Aspects of the thermal biology of the Texas tortoise, *Gopherus berlandieri* (Reptilia, Testudines, Testudinidae)." *Journal of Herpetology* 11:147–53.

———. 1983. "Population structure, density, and movements of the Texas tortoise, *Gopherus berlandieri*." *Southwestern Naturalist* 28:387–98.

———. 1989. "Egg production by the Texas tortoise, *Gopherus berlandieri*, in southern Texas." *Copeia* 1989:588–96.

———. 2000. "Conservation status of the Texas tortoise *Gopherus berlandieri*." Occasional Papers, Museum of Texas Tech University No. 196. Lubbock: Texas Tech University.

Kazmaier, R. T., E. C. Hellgren, and D. C. Ruthven, III. 2001a. "Habitat selection by the Texas Tortoise in a managed thornscrub ecosystem." *Journal of Wildlife Management* 65:653–60.

Kazmaier, R. T., E. C. Hellgren, and D. C. Ruthven, III. 2002. "Home range and dispersal of Texas tortoises, *Gopherus berlandieri*, in a managed thornscrub ecosystem." *Chelonian Conservation and Biology* 4:488–96.

Kazmaier, R. T., E. C. Hellgren, D. C. Ruthven, III, and D. R. Synatzske. 2001b. "Effects of grazing on the demography and growth of the Texas tortoise." *Conservation Biology* 15: 1091–101.

Kazmaier, R. T., E. C. Hellgren, D. R. Synatzske, and J. C. Rutledge. 2001c. "Mark-recapture analysis of population parameters in a Texas tortoise (*Gopherus berlandieri*) population in southern Texas." *Journal of Herpetology* 35:410–17.

Killebrew, F. C., and R. R. McKown. 1978. "Mitotic chromosomes of *Gopherus berlandieri* and *Kinixys belliana belliana* (Testudines, Testudinidae)." *Southwestern Naturalist* 23:162–64.

Kjos, S. A., K. F. Snowden, and J. K. Olson. 2009. "Biogeography and *Trypanosoma cruzi* infection prevalence of Chagas disease vectors in Texas, USA." *Vector-Borne and Zoonotic Diseases* 9:41–49.

Krebs, C. J. 2001. *Ecology: The Experimental Analysis of Distribution and Abundance*, 5th ed. San Francisco: Benjamin Cummings.

Krohne, D. T. 2001. *General Ecology*, 2nd ed. Pacific Grove, Calif.: Brooks/Cole.

Lamb, T., J. C. Avise, and J. W. Gibbons. 1989. "Phylogeographical patterns of mitochondrial DNA of the desert tortoise (*Xerobates agassizi*), and evolutionary relationships among North American gopher tortoises." *Evolution* 43:76–87.

Lamb, T., and C. Lydeard. 1994. "A molecular phylogeny of the gopher tortoises, with comments on familial relationships within the Testudinoidea." *Molecular Phylogenetics and Evolution* 3:283–91.

Landers, J. L., J. A. Garner, and W. A. McRae. 1980. "Reproduction of gopher tortoises (*Gopherus polyphemus*) in southwestern Georgia." *Herpetologica* 36:353–61.

Legler, J. M. 1959. "A new tortoise, genus *Gopherus*, from north-central Mexico." University of Kansas Publications, Museum of Natural History 11:335–43.

———. 1962. "The os transiliens in two species of tortoises, genus *Gopherus*." *Herpetologica* 18:68–69.

Li, C., X. Wu, O. Rieppel, L. Wang, and L. Zhao. 2008. "An ancestral turtle from the late Triassic of southwestern China." *Nature* 456:497–501.

Licht, P. 1972. "Environmental physiology of reptilian breeding cycles: Role of temperature." *General Comparative Endocrinology*, supplement no. 3:447–88.

Lonard, R. I., and F. W. Judd. 1985. "Effects of severe freeze on native woody plants in the lower Rio Grande Valley, Texas." *Southwestern Naturalist* 30:397–403.

———. 1991. "Comparison of the effects of severe freezes of 1983 and 1989 on native woody plants in the lower Rio Grande Valley, Texas." *Southwestern Naturalist* 36:213–17.

Long, D. R., and F. L. Rose. 1989. "Pelvic girdle size relationships in three turtle species." *Journal of Herpetology* 23:315–18.

Lowe, C. H., P. J. Lardner, and E. A. Halpern. 1971. "Supercooling in reptiles and other vertebrates." *Comparative Biochemistry and Physiology* 39A:125–35.

Luckenbach, R. A. 1982. "Ecology and management of the desert tortoise (*Gopherus agassizii*) in California." In *North American Tortoises: Conservation and Ecology*, edited by R. B. Bury, 1–37. U. S. Fish and Wildlife Service, Wildlife Research Report 12.

Mahmoud, I. Y. 1968. "Nesting behavior in the western painted turtle, *Chrysemys picta belli*." *Herpetologica* 24:158–62.

Mann, C. C. 2011. *"1493: Uncovering the New World Columbus Created."* New York: Knopf Doubleday.

Mares, M. A. 1971. "Coprophagy in the Texas tortoise, *Gopherus berlandieri*." *Texas Journal of Science* 23:300.

Martin, J. H., and L. W. McEachron. 1996. "Historical annotated review of winter kills of marine organisms in Texas bays." Coastal Fisheries Branch, Management Data Series no. 50. Austin: Texas Parks and Wildlife Department.

Mayr, E. 1970. *Populations, Species, and Evolution*. Cambridge, Mass.: Belknap Press of Harvard University Press.

McAllister, C. T., R. Bursey, and S. E. Trauth. 2008. "New host and geographic distribution records for some endoparasites (Myxosporea, Trematoda, Cestoidea, Nematoda) of amphibians and reptiles from Arkansas and Texas, U.S.A." *Comparative Parasitology* 75:241–54.

McCord, R. D. 2002. "Fossil history and evolution of gopher tortoises (genus *Gopherus*)." In *The Sonoran Desert Tortoise*, edited by T. R. Van Devender, 52–66. Tucson: University of Arizona Press.

McCutcheon, F. H. 1943. "The respiratory mechanism of turtles." *Physiological Zoology* 16:255–69.

McKeown, S., J. O. Juvik, and D. E. Meir. 1982. "Observations on the reproductive biology of *Geochelone emys* and *Geochelone yniphora* in the Honolulu Zoo." *Zoo Biology* 1:223–35.

McLendon, T. 1991. "Preliminary description of the vegetation of South Texas exclusive of coastal saline zones." *Texas Journal of Science* 43:13–32.

McMahan, C. A., R. G. Frye, and K. L. Brown. 1984. "The vegetation types of Texas including cropland." Austin: Texas Parks and Wildlife Department, Wildlife Division.

McRae, W. A., J. L. Landers, and G. D. Cleveland. 1981. "Sexual dimorphism in the gopher tortoise (*Gopherus polyphemus*)." *Herpetologica* 37:46–52.

Metcalf, E., and A. L. Metcalf. 1970. "Observations on ornate box turtles (*Terrapene ornata ornata* Agassiz)." *Transactions of the Kansas Academy of Science* 73:95–117.

Mittleman, M. B., and B. C. Brown. 1947. "Notes on *Gopherus berlandieri* (Agassiz)." *Copeia* 1947:211.

Moll, E. O. 1979. "Reproductive cycles and adaptations." In *Turtles: Perspectives and Research*, edited by M. Harless and H. Morlock, 305–31. New York: John Wiley & Sons.

Moon, J. C., E. D. McCoy, H. R. Mushinsky, and S. A. Karl. 2006. "Multiple paternity and breeding system in the gopher tortoise, *Gopherus polyphemus*." *Journal of Heredity* 97:150–57.

Morafka, D. J. 1982. "The status and distribution of the Bolson tortoise (*Gopherus flavomarginatus*)." In *North American Tortoises*, edited by R. B. Bury, 71–94. U.S. Fish and Wildlife Service, Wildlife Research Report 12.

Morafka, D. J., G. Aguirre L., and R. W. Murphy. 1994. "Allozyme differentiation among gopher tortoises (*Gopherus*): Conservation genetics and phylogenetic and taxonomic implications." *Canadian Journal of Zoology* 72:1665–71.

Morafka, D. J., and K. H. Berry. 2002. "Is *Gopherus agassizii* a desert-adapted tortoise, or an exaptive opportunist? Implications for tortoise conservation." *Chelonian Conservation and Biology* 4:263–87.

Murphy, R. W., K. H. Berry, T. Edwards, A. L. Leviton, A. Lathrop, and J. D. Riedle. 2011. "The dazed and confused identity of Agassiz's land tortoise, *Gopherus agassizii* (Testudines, Testudinidae) with the description of a new species, and its consequences for conservation." *ZooKeys* 113:39–71.

Mushinsky, H. R., D. S. Wilson, and E. D. McCoy. 1994. "Growth and sexual dimorphism of *Gopherus polyphemus* in central Florida." *Herpetologica* 50:119–28.

Mutz, J. L., C. J. Scifres, D. L. Drawe, T. W. Box, and R. E. Whitson. 1978. "Range vegetation after mechanical brush treatment on the coastal prairie." Texas Agricultural Experiment Station Contribution No. B-1191.

Nagashima, H., F. Sugahara, M. Takechi, R. Ericsson, Y. Kawashima-Ohya, Y. Narita, and S. Kuratani. 2009. "Evolution of the turtle body plan by the folding and creation of new muscle connections." *Science* 325:193–96.

Nagy, K. A., M. W. Tuma, and L. S. Hillard. 2011. "Shell hardness measurement in juvenile desert tortoises." *Herpetological Review* 42:191–95.

Neck, R. W. 1977. "Cutaneous myiasis in *Gopherus berlandieri* (Reptilia, Testudines, Testudinidae)." *Journal of Herpetology* 11:96–98.

Newcomb, W. W., Jr. 1980. *The Indians of Texas from Prehistoric to Modern Times.* Austin: University of Texas Press.

Ohlendorf, S. M., J. M. Bigelow, and M. M. Standifer, trans. 1980. *Journey to Mexico during the Years 1826 to 1834.* Texas State Historical Association in cooperation with the Center for Studies in Texas History, University of Texas, Austin.

Olson, R. E. 1976. "Weight regimes in the Texas tortoise *Gopherus berlandieri*." *Texas Journal of Science* 17:321–23.

———. 1987. "Evaporative water loss in the tortoise *Gopherus berlandieri* in ambient temperature regimes." *Bulletin of the Maryland Herpetological Society* 23:93–100.

———. 1989. "Notes on evaporative water loss in terrestrial chelonians. *Bulletin of the Maryland Herpetological Society* 25:49–57.

Ottley, J. R., and V. M. Velázques Solis. 1989. "An extant, indigenous tortoise population in Baja California Sur, Mexico, with the description of a new species of *Xerobates* (Testudines: Testudinidae)." *Great Basin Naturalist* 49:496–502.

Patterson, R. 1971a. "The os transiliens in four species of tortoises, genus *Gopherus*." *Bulletin of the Southern California Academy of Science* 72:51–52.

———. 1971b. "The role of egg urination in egg predator defense in the desert tortoise (*Gopherus agassizii*)." *Herpetologica* 27:197–99.

———. 1971c. "Visual cliff perception in tortoises." *Herpetologica* 27:339–41.

Patterson, R., and B. Brattstrom. 1972. "Growth of captive *Gopherus agassizii*." *Herpetologica* 28:169–71.

Patterson, W. C. 1966. "Hearing in the turtle." *Journal of Auditory Research* 6:453–64.

Paxson, D. W. 1961. "An observation of eggs in a tortoise shell." *Herpetologica* 17:278–79.

Pérez, J. C., S. Pichyangkul, and V. E. Garcia. 1979. "The resistance of three species of warm-blooded animals to western diamondback rattlesnake (*Crotalus atrox*) venom." *Toxicon* 17:601–607.

Pianka, E. R. 2000. *Evolutionary Ecology*, 6th ed. New York: Addison Wesley Longman.

Pope, C. H. 1939. *Turtles of the United States and Canada.* New York: A. A. Knopf.

Raun, G. G. 1966. "A population of wood rats (*Neotoma microceps*) in southern Texas." *Bulletin of the Texas Memorial Museum* 11:1–62.

Ray, C. E. 1945. "A sesamoid bone in jaw musculature of *Gopherus polyphemus* (Reptilia: Testudinidae)." *Anatomischer Anzeiger* 107 (1/5): 85–91.

Reynoso, V. H., and M. Montellano-Ballesteros. 2004. "A new giant turtle of the genus

Gopherus (Chelonia: Testudinidae) from the Pleistocene of Tamaulipas, Mexico, and a review of the phylogeny and biogeography of gopher tortoises." *Journal of Vertebrate Paleontology* 24:822–37.

Richmond, N. D. 1964. "The mechanical function of the testudinate plastron." *American Midland Naturalist* 72:50–56.

Rieppel, O. 2009. "How did the turtle get its shell?" *Science* 325:154–55.

Rose, F. L. 1970. "Tortoise chin gland fatty acid composition: Behavioral significance." *Comparative Biochemistry and Physiology* 32:577–80.

———. 1983. "Aspects of the thermal biology of the Bolson tortoise, *Gopherus flavomarginatus*." Occasional Papers, Museum of Texas Tech University No. 89:1–8.

———. 1986. "Carapace regeneration in *Terrapene carolina* (Chelonia: Testudinidae)." *Southwestern Naturalist* 31:131–34.

Rose, F. L., R. Drotman, and W. G. Weaver. 1969. "Electrophoresis of chin gland extracts of *Gopherus* (tortoises)." *Comparative Biochemistry and Physiology* 29:847–51.

Rose, F. L., and F. W. Judd. 1977. "Aspects of the thermal biology of the Texas tortoise, *Gopherus berlandieri* (Reptilia, Testudines, Testudinidae)." *Journal of Herpetology* 11:147–53.

———. 1982. "Biology and status of Berlandier's tortoise (*Gopherus berlandieri*)." In *North American Tortoises: Conservation and Ecology*, edited by R. B. Bury, 57–70. U.S. Fish and Wildlife Service, Wildlife Research Report 12.

———. 1989. "*Gopherus berlandieri*." In *The Conservation Biology of Tortoises*, edited by I. R. Swingland and M. W. Klemens, 8–9. Occasional Papers of the IUCN Species Survival Commission No. 5.

———. 1991. "Egg size versus carapace-xiphiplastron aperture size in *Gopherus berlandieri*." *Journal of Herpetology* 25:248–50.

Rose, F. L., F. W. Judd, and M. F. Small. 2011. "Survivorship of two coastal populations of *Gopherus berlandieri*." *Journal of Herpetology* 45:75–78.

Rose, F. L., J. R. Koke, R. Koehn, and D. Smith. 2001. "Identification of the etiological agent for necrotizing scute disease in the Texas tortoise." *Journal of Wildlife Diseases* 37:223–28.

Rose, F. L., M. E. T. Scioli, and M. P. Moulton. 1988. "Thermal preferentia of Berlandier's tortoise (*Gopherus berlandieri*) and the ornate box turtle (*Terrapene ornata*)." *Southwestern Naturalist* 33:357–61.

Scalise, J. L. 2010. "Food habits and selective foraging by the Texas tortoise (*Gopherus berlandieri*)." MS thesis, Texas State University.

Schad, G. A., R. Knowles, and E. Merrovitch. 1964. "The occurrence of *Lampropedia* in the intestines of some reptiles and nematodes." *Canadian Journal of Microbiology* 10:801–804.

Schwagmeyer, P., and S. Woontner. 1986. "Scramble competition polygyny in thirteen-lined ground squirrels: The relative contributions of overt conflict and competitive mate searching." *Behavioral Ecology and Sociobiology* 19:359–64.

Selander, R. K., R. F. Johnston, B. J. Wilks, and G. G. Raun. 1962. "Vertebrates from the barrier islands of Tamaulipas, Mexico." University of Kansas Museum of Natural History Publications 12:309–45.

Silverman, S., and D. L. Janssen. 1996. "Diagnostic imaging." In *Reptile Medicine and Surgery*, edited by D. R. Mader. Philadelphia: W. B. Saunders.

Smith, H. 1958. "Total regeneration of the carapace in the box turtle." *Turtox News* 36:234–37.

Smith, H. M., and S. O. Brown. 1946. "A hitherto neglected integumentary gland in the Texas tortoise." Abstract. *Proceedings and Transactions of the Texas Academy of Science* 30:59.

Smith, L. L. 1992. "Nesting ecology, female home range and activity patterns, and hatchling survivorship in the gopher tortoise (*Gopherus polyphemus*)." MS thesis, University of Florida, Gainesville.

Snipes, K. P., and E. L. Biberstein.1982. "*Pasteurella testudinis* sp. nov.: A parasite of desert tortoises (*Gopherus agassizii*)." *International Journal of Systematic Bacteriology* 32:201–10.

Spray, D. C., and M. L. May. 1972. "Heating and cooling rates in four species of turtles." *Comparative Biochemistry and Physiology* 43A:507–522.

Steinbeck, J. 1995. *Log from the Sea of Cortez*. New York: Penguin Group USA.

Strecker, J. K. 1915. "Reptiles and amphibians of Texas." *Baylor University Bulletin* 18:1–82.

———. 1928. "The eggs of *Gopherus berlandieri* Agassiz." *Contributions from Baylor University Museum* 18:6.

Stubbs, D., I. R. Swingland, A. Hailey, and E. Pulford. 1985. "The ecology of the Mediterranean tortoise, *Testudo hermanni*, in northern Greece (The effects of catastrophe on population structure and density)." *Biological Conservation* 31:125–52.

Tanzer, E. C., E. O. Morrison, and C. Hoffpauir. 1966. "New locality records for amphibians and reptiles in Texas." *Southwestern Naturalist* 11:131–32.

Tinkle, D. W., J. D. Congdon, and P. C. Rosen. 1981. "Nesting frequency and success: Implication for the demography of painted turtles." *Ecology* 62:1426–32.

Tremblay, T. A., W. A. White, and J. A. Raney. 2005. "Native woodland loss during the mid 1990s in Cameron County, Texas." *Southwestern Naturalist* 50:479–82.

Tristan, T. 2009. "Seroprevalence of *Mycoplasma agassizii* in wild caught and rescued Texas tortoises (*Gopherus berlandieri*) in South Texas." *Journal of Herpetological Medicine and Surgery* 19:115–18.

True, F. W. 1881. "On North American land tortoises of the genus *Xerobates*." *Proceedings of the U.S. National Museum* 4:434–49.

Tunnell, J. W., Jr. 2002. "Geography, climate, and hydrology." In *The Laguna Madre of Texas and Tamaulipas*, edited by J. W. Tunnell, Jr. and F. W. Judd, 7–27. College Station: Texas A&M University Press.

Turner, F. B., P. Hayden, B. L. Burge, and J. B. Roberson. 1986. "Egg production by a desert tortoise (*Gopherus agassizii*) in California." *Herpetologica* 41:93–104.

Vecchio, S. D., R. L. Burke, L. Rugiero, M. Capula, and L. Luiselli. 2011. "Seasonal changes in the diet of *Testudo hermanni hermanni* in central Italy." *Herpetologica* 67:236–49.

Viosca, P. 1927. "Note on *Gopherus berlandieri* in Louisiana." *Copeia* 1927:83–84.

Voigt, W. G. 1975. "Heating and cooling rates and their effects upon heart rate and subcutaneous temperatures in the desert tortoise, *Gopherus berlandieri*." *Comparative Biochemistry and Physiology* 52A:527–31.

Voigt, W. G., and C. R. Johnson. 1976. "Aestivation and thermoregulation in the Texas tortoise, *Gopherus berlandieri*." *Comparative Biochemistry and Physiology* 53A:41–44.

———. 1977. "Physiological control of heat exchange rates in the Texas tortoise, *Gopherus berlandieri*." *Comparative Biochemistry and Physiology* 56A:495–98.

Weathers, W. W., and F. N. White. 1971. "Physiological thermoregulation in turtles." *American Journal of Physiology* 221:704–10.

Weaver, W. G., Jr. 1970. "Courtship and combat behavior in *Gopherus berlandieri*." *Bulletin of the Florida State Museum* 15:1–43.

Wells, K. D. 1977. "The social behaviour of anuran amphibians." *Animal Behaviour* 25:666–93.

Westhouse, R. A., E. R. Jacobson, R. K. Harris, K. R. Winter, and B. L. Homer. 1996.

"Respiratory and pharyngo-esophageal iridovirus infection in a gopher tortoise (*Gopherus polyphemus*)." *Journal of Wildlife Diseases* 31:682–86.

Whittaker, R. H. 1975. *Communities and Ecosystems*, 2nd ed. New York: Macmillan.

Wilbur, H. M. 1975. "The evolutionary and mathematical demography of the turtle, *Chrysemys picta*." *Ecology* 56:64–77.

Winokur, R. M., and J. M. Legler. 1975. "Chelonian mental glands." *Journal of Morphology* 147:275–92.

Woodbury, A. M. 1952. "Hybrids of *Gopherus berlandieri* and *G. agassizii*." *Herpetologica* 8:33–36.

Woodbury, A. M., and R. Hardy. 1948. "Studies of the desert tortoise, *Gopherus agassizii*." *Ecological Monographs* 18:145–200.

Zangerl, R. 1939. "The homology of the shell elements in turtles." *Journal of Morphology* 65:383–406.

———. 1969. "The turtle shell." In *Morphology A*. Vol. 1, *Biology of the Reptilia*, edited by C. Gans, A. d'A. Bellairs, and T. S. Parsons, 311–39. London: Academic Press.

Zug, G. R. 1966. "The penial morphology and the relationships of cryptodiran turtles." *Occasional Papers of the Museum of Zoology, University of Michigan*, no. 6467:1–24.

———. 1993. *Herpetology: An Introductory Biology of Amphibians and Reptiles*. San Diego: Academic Press.

Zug, G. R., L. J. Vitt, and J. P. Caldwell. 2001. *Herpetology: An Introductory Biology of Amphibians and Reptiles*, 2nd ed. San Diego: Academic Press.

Index

Acacia: *berlandieri*, 32, 165; *minuata*, 32, 165; *rigidula*, 32, 165; *schaffneri*, 165
Activity, 71, 74, 90, 96, 101, 103–107, 109, 115, 117–18, 129–30, 134, 152
Agassiz, Louis, 4, 6–7
Agassiz desert tortoise. See *Gopherus*: *agassizii*
Age, 134
Age structure, 132, 134
Albumen, 73–74, 76, 78–79, 114
Alfisols, 29
Allometry, 14
Amerindians, 150
Anacacho Mountains, Tex., 29
Anacahuita. See *Cordia boissieri*
Antilocapra americana, 48
Aridisols, 29
Atlas, 55
Auffenberg, Walter, 71
Axis, 55

Baja California Sur, Mexico, 17
Barrier islands, 26, 29, 33, 96
Baytril (Enrofloxacin), 163
Berlandier, Jean Louis, 3–4, 6–8
Betadine, 162
Biomes, 30
Biotic province, 24, 26–27, 29–30, 165
Bison. See *Bos bison*
Blackbrush. See *Acacia*: *rigidula*
Bladder stone, 43
Bluestem, little. See *Schizachyrium*: *scoparium*

Bobcat. See *Lynx rufus*
Bolson tortoise. See *Gopherus*: *flavomarginatus*
Border Paloverde. See *Parkinsonia*: *texana* var. *macra*
Borrichia frutescens, 36
Bos bison, 48
Box turtle. See *Terrapene ornata*
Brownsville, Tex., 27–29, 103
Buchloë dactyloides, 36
Buffalo grass. See *Buchloë dactyloides*
Buffelgrass. See *Pennisetum ciliare*
Burrowing, 30, 88, 105–107

Caecum, 61
Calcium, 37–38, 43, 46, 50, 75–76, 86, 113–14, 133, 160
Canis latrans, 38, 154
Captive propagation, 159
Caracara, crested. See *Caracara cheriway*
Caracara cheriway, 38
Carapace, 8, 16–18, 20–22, 38–39, 43–45, 48–50, 52, 55–58, 62, 71, 76, 78–79, 82–86, 88, 91–92, 103, 110, 112–14, 133–35, 142, 146, 151, 158, 160, 166
Carapace bones: costals, 50; neurals, 50, 55; nuchal, 50, 92; peripherals, 39, 50; pygal, 50, 78; suprapygal, 50
Carapace scutes: marginal, 50, 91; nuchal, 50; pleural, 50; vertebral, 50
Caruncle (egg tooth), 80, 81
Casto, Stanley, 40
Ceftazidime, 163

p. iii: Large male Texas tortoise from Cameron County, Texas (see p. 121, plate 25).

Copyedited by Laurel Anderton
Design and composition by Chris Crochetière, BW&A Books, Inc.
Set in Minion and The Sans
Jacket design by Tony Roberts
Image prepress by University of Oklahoma Printing Services
Color gallery printed by John P. Pow Company
Text printed and bound by Edwards Brothers Malloy